Gas
Chromatography

ANALYTICAL CHEMISTRY BY OPEN LEARNING

ACOL (Analytical Chemistry by Open Learning) is a well established series which comprises 33 open learning books and 8 computer based training packages. This open learning material covers all the important techniques and fundamental principles of analytical chemistry.

Books

Samples and Standards
Sample Pretreatment
Classical Methods Vols I and II
Measurement, Statistics and Computation
Using Literature
Instrumentation
Chromatographic Separations
Gas Chromatography
High Performance Liquid Chromatography
Electrophoresis
Thin Layer Chromatography
Visible and Ultraviolet Spectroscopy
Fluorescence and Phosphorescence
Infrared Spectroscopy
Atomic Absorption and Emission
    Spectroscopy
Nuclear Magnetic Resonance
    Spectroscopy

X-Ray Methods
Mass Spectrometry
Scanning Electron Microscopy and
    Microanalysis
Principles of Electroanalytical Methods
Potentiometry and Ion Selective Electrodes
Polarography and Voltammetric Methods
Radiochemical Methods
Clinical Specimens
Diagnostic Enzymology
Quantitative Bioassay
Assessment and Control of Biochemical
    Methods
Thermal Methods
Microprocessor Applications
Chemometrics
Environmental Analysis
Quality in the Analytical Chemistry
    Laboratory

Software

Atomic Absorption Spectroscopy
High Performance Liquid Chromatography
Polarography
Radiochemistry
Gas Chromatography
Fluorescence
Quantitative IR–UV
Chromatography

Series Coordinator: David J. Ando

Further information: ACOL Office
                        Greenwich University Press
                        Unit 42, Dartford Trade Park
                        Hawley Road
                        Dartford
                        DA1 1PF

Gas Chromatography

Analytical Chemistry by Open Learning

Second Edition

Author:
IAN A. FOWLIS
University of Northumbria at Newcastle

Published on behalf of ACOL (University of Greenwich)
by
JOHN WILEY & SONS
Chichester • New York • Brisbane • Toronto • Singapore

Copyright © 1995 University of Greenwich, UK

Published by John Wiley & Sons Ltd.
            Baffins Lane, Chichester,
            West Sussex PO19 1UD, England
            Telephone National   Chichester (01243) 779777
                     International +44 1243 779777

Other Wiley Editorial Offices

John Wiley & Sons, Inc, 605 Third Avenue,
New York, NY 10158-0012, USA

Jacaranda Wiley Ltd, 33 Park Road, Milton,
Queensland 4064, Australia

John Wiley & Sons (Canada) Ltd, 22 Worcester Road,
Rexdale, Ontario M9W 1LJ, Canada

John Wiley & Sons (SEA) Pte Ltd, 37 Jalan Pemimpin #05-04,
Block B, Union Industrial Building, Singapore 2057

Library of Congress Cataloging-in-Publication Data

Fowlis, Ian A.
    Gas chromatography : analytical chemistry by open learning / Ian
A. Fowlis. — 2nd ed.
        p.   cm. — (Analytical Chemistry by Open Learning)
    'Published on behalf of ACOL (University of Greenwich).'
    Includes bibliographical references and index.
    ISBN 0-471-95467-5 : — ISBN 0-471-95468-3 (pbk.)

    1. Gas chromatography — Programmed instruction. 2. Chemistry,
Analytic — Programmed instruction.   I. ACOL (Project)   II. Title.
III. Series: Analytical Chemistry by Open Learning (Series)
QD79. C45F68   1994
543' 0896'077 — dc20   94-35413
                                                               CIP

British Library Cataloguing in Publication Data

A catalogue record for this book is available from the British Library

ISBN 0 471 95467 5 (cloth)
ISBN 0 471 95468 3 (paper)

Typeset in 11/13pt Times by Mackreth Media Services, Hemel Hempstead
Printed and bound in Great Britain by Biddles Ltd, Guildford, Surrey

THE ACOL PROJECT

This series of easy to read books has been written by some of the foremost lecturers in Analytical Chemistry in the United Kingdom. These books are designed for training, continuing education and updating of all technical staff concerned with Analytical Chemistry.

These books are for those interested in Analytical Chemistry and instrumental techniques who wish to study in a more flexible way than traditional institute attendance or to augment such attendance.

ACOL also supply a range of training packages which contain computer software together with the relevant ACOL book(s). The software teaches competence in the laboratory by providing experience of decision making in the laboratory often based on the simulation of instrumental output while the books cover the requisite underpinning knowledge.

The Royal Society of Chemistry used ACOL material to run a regular series of courses based on distance learning and regular workshops.

Further information on all ACOL materials and courses may be obtained from:

ACOL Office, Greenwich University Press, Unit 42, Dartford Trade Park, Hawley Road, Dartford DA1 1PF, Tel: 0181-331 9600 Fax: 0181-331 9672.

How to Use an Open Learning Book

Open learning books are designed as a convenient and flexible way of studying for people who, for a variety of reasons, cannot use conventional education courses. You will learn from this book the principles of one subject in Analytical Chemistry, but only by putting this knowledge into practice, under professional supervision, will you gain a full understanding of the analytical techniques described.

To achieve the full benefit from an open learning book you need to plan your place and time of study.

- Find the most suitable place to study where you can work without disturbance.

- If you have a tutor supervising your study discuss with him, or her, the date by which you should have completed this text.

- Some people study perfectly well in irregular bursts, however most students find that setting aside a certain number of hours each day is the most satisfactory method. It is for you to decide which pattern of study suits you best.

- If you decide to study for several hours at once, take short breaks of five or ten minutes every half hour or so. You will find that this method maintains a higher overall level of concentration.

Before you begin a detailed reading of the book, familiarise yourself with the general layout of the material. Have a look at the course contents list at the front of the book and flip through the pages to get a general impression of the way the subject is dealt with. You will find that there is space on the pages to make comments alongside the text as you study — your own notes for highlighting points that you feel are particularly important. Indicate in the margin the points you would like to discuss further with a tutor or fellow student. When you come to revise, these personal study notes will be very useful.

∏ When you find a paragraph in the text marked with a symbol such as is shown here, this is where you get involved. At this point you are directed to do things: draw graphs, answer questions, perform calculations, etc. Do make an attempt at these activities. If necessary cover the succeeding response with a piece of paper until you are ready to read on. This is an opportunity for you to learn by participating in the subject and although the text continues by discussing your response, there is no better way to learn than by working things out for yourself.

We have introduced self-assessment questions (SAQs) at appropriate places in the text. These SAQs provide for you a way of finding out if you understand what you have just been studying. There is space on the page for your answer and for any comments you want to add after reading the author's response. You will find the author's response to each SAQ at the end of the book. Compare what you have written with the response provided and read the discussion and advice.

At intervals in the book you will find Lists of Learning Objectives. These will give you a check-list of tasks you should then be able to achieve at these various stages of the Unit..

You can revise the Unit, perhaps for a formal examination, by re-reading the Objectives, and by working through some of the SAQs. This should quickly alert you to areas of the text that need further study.

At the end of this book you will find, for reference, lists of commonly used scientific symbols and values, units of measurement and also a periodic table.

Contents

Study Guide

This Unit is intended to provide you with a working knowledge of gas chromatography. It will not turn you into a fully experienced gas chromatographer — only months and years of practice can do that, for gas chromatography is still something of an art. Many of us have learned through standing in front of a gas chromatograph and making endless mistakes. It is very frustrating, but it is probably the best way to learn to recognise the symptoms of the things which can go wrong with gas chromatography. There are many settings on a gas chromatograph which can be wrongly chosen, and many physical malfunctions that have nothing to do with the chemistry of the process but affect the outcome of an analysis dramatically. They have to be experienced, or at least described to you before you learn to recognise the symptoms and diagnose the faults. This Unit will try to provide you with the equivalent of that experience by adopting the standpoint of a practising gas chromatographer.

Even so, while emphasising the practical side of the subject, the theoretical side will not be ignored. It is only by understanding the processes occurring in your instrument that you will get the best out of it. The purpose of considering the theory, though, is quite clearly to improve your practical performance.

It will be assumed that you have an understanding of chemistry equivalent to that of a student who has passed HNC or HTC in chemistry (BTEC), and a knowledge of physics up to at least GCE (OL). You may also have some basic understanding of chromatography.

You may find that another author's views on gas chromatography will clarify, for you, some aspects of this text. A number of recent and not-so-recent texts and publications are listed in the Bibliography, some of which are included for their historical interest.

Inevitably, in a distance learning package you will not get anything like enough 'hands-on' practical experience, but at your regional centre you may have the opportunity to use gas chromatographs and computer controlled simulations which will help you along the road to becoming an experienced gas chromatographer.

Supporting Practical Work

1. GENERAL CONSIDERATIONS

Gas chromatographs are one of the workhorses of many laboratories so it is quite likely that you will be able to gain practical experience at your place of work.

In general, gas chromatographers are a friendly bunch who will be only too glad to show you the equipment they use and to suggest experiments you may try out on the instruments available. Remember of course that you will have to slot in with the priorities of the laboratory and do not neglect your own specific duties in your enthusiasm for your new interest.

Take note carefully of the instructions given to you and be prepared to discuss various aspects of this text as you progress with the chosen exercises. These discussions will help to reinforce your understanding of the text and its relationship to the outcome of your practical experiments.

It will probably be of more value to you if you obtain access to an older instrument in the first instance since you can then become thoroughly familiar with the hardware, changing columns and the controls rather than just addressing a computer which controls the system. Gas chromatographs are very robust and provided you handle syringes with care as instructed you are unlikely to do any damage. Having mastered the basics, it is then essential to gain experience using both packed and capillary columns. If, in addition, you can spend some time running high resolution gas chromatography (HRGC)–mass spectrometry (MS) or HRGC–infrared (IR) spectroscopy then so much the better.

Finally, always endeavour to produce chromatograms which look good. Gas chromatography is an art and usually if the chromatogram looks good it is good in terms of separation and sensitivity.

2. AIMS

(*a*) To provide a basic experience of using gas chromatographic equipment.

(*b*) To illustrate the effect on separation in gas chromatography of the various operating parameters using both packed and capillary columns.

(*c*) To illustrate the feasibility of qualitative and quantitative analysis.

(*d*) To illustrate how gas chromatography can be combined with spectroscopic methods to provide identification of eluted components.

3. SUGGESTED EXPERIMENTS

(*a*) The examination of the effect of stationary phase, column length, temperature, carrier gas flow-rate, sample size, etc., upon the separation of benzene, cyclohexane and ethanol.

(*b*) Comparison of the resolving power of packed and capillary columns on a mixture of long-chain fatty acid esters.

(*c*) The determination of the concentration of ethanol in a dilute aqueous sample in the concentration range relevant to drink–drive legislation (approximately 100 mg/100 ml) using propan-1-ol as internal standard.

(*d*) Optimise conditions for the analysis of spearmint oil by capillary column gas chromatography.

(*e*) Compare the responses obtained for a sample mixture containing toluene, *p*-chlorophenol, bromobenzene, dibromobenzene and n-butanol using flame ionisation, electron capture and mass spectrometry detection.

Bibliography

(1) P. A. Sewell and B. Clark, *ACOL: Chromatographic Separations*, Wiley, 1987.

(2) H. H. Willard, L. L. Merritt, J. A. Dean and F. A. Settle, *Instrumental Methods of Analysis* (7th Edn), Wadsworth Publishing, 1988.

(3) G. Schomburg, *Gas Chromatography; A Practical Course*, VCH, 1990.

(4) S. E. Manahan, *Quantitative Chemical Analysis*, Brooks Cole, 1986.

(5) M. J. E. Golay, *Gas Chromatography 1960* (Ed. R. P. W. Scott), Butterworths, 1960.

(6) Kurt Grob, *Making and Manipulating Capillary Columns for Gas Chromatography*, Dr Alfred Heuthig GmbH, 1986.

(7) Konrad Grob, *Classical Split and Splitless Injection in Capillary Gas Chromatography*, Dr Alfred Heuthig GmbH, 1988.

(8) Konrad Grob, *On-Column Injection in Capillary Gas Chromatography*, Dr Alfred Heuthig GmbH, 1987.

(9) R. Davis and M. Frearson, *ACOL: Mass Spectrometry*, Wiley, 1987.

(10) R. M. Silverstein, G. C. Bassler and T. C. Morrill, *Spectrometric Identification of Organic Compounds* (5th Edn), Wiley, 1991.

(11) W. George and P. McIntyre, *ACOL: Infrared Spectroscopy*, Wiley, 1987.

(12) K. Blau and J. M. Halket, *Handbook of Derivatives for Chromatography*, Wiley, 1993.

(13) D. Welti, *Infrared Vapour Spectra*, Heyden/Sadtler, 1980.

(14) H. Hachenberg and A. P. Schmidt, *Gas Chromatography Head Space Analysis*, Heyden, 1977.

(15) B. Kolb, *Applied Headspace Gas Chromatography*, Heyden, 1980.

(16) A. T. James and A. J. P. Martin, *Biochem. J.*, **50**, 679 (1952).

(17) I. A. Fowlis, R. J. Maggs and R. P. W. Scott, *J. Chromatogr.*, **15**, 471 (1964).

(18) L. S. Ettre, *J. High Resolut. Chromatogr., Chromatogr. Commun.*, **10**, 221 (1987).

(19) D. Rood, *Quantitative Analysis using Chromatographic Techniques* (Separation Science Series), (Ed. E. Katz), Wiley, 1988.

(20) W. G. Jennings, *Gas Chromatography with Glass Capillary Columns*, Academic Press, 1978.

(21) C. F. Poole and S. K. Poole, *Chromatogr. Today*, Elsevier, 1991.

(22) J. Q. Walker, M. T. Jackson Jr, and J. B. Maynard, *Chromatographic Systems, Maintenance and Troubleshooting of Gas Chromatography and Liquid Chromatography* (2nd Edn), Academic Press, 1985.

(23) M. S. Klee, *GC Inlets — An Introduction*, Hewlett-Packard, 1990.

(24) R. Buffington and M. Wilson, *Detectors for Gas Chromatography — A Practical Primer*, Hewlett-Packard, 1987.

(25) K. J. Hyver (Ed.), *High Resolution Gas Chromatography* (3rd Edn), Hewlett-Packard, 1989.

(26) B. V. Loffe and A. G. Vitenberg, *Headspace Analysis and Related Methods in Gas Chromatography*, Wiley–Interscience, 1984.

Acknowledgements

Figures 5.2a, 5.3a, 5.5a, 5.6b and 6.2a are taken from instrument catalogues published by Fisons Instruments SpA, Rodano, Italy and are reproduced with permission.

Figure 5.2b is taken from an instrument catalogue published by the Hewlett-Packard Company, and is reproduced with permission. The copyright remains with this company.

Figures 6.2b and 6.3a are taken from instrument catalogues published by The Perkin-Elmer Corporation and are reproduced with permission.

Figure 10.5a is taken from an instrument catalogue published by Chrompack (UK) Ltd and is reproduced with permission.

1. Introduction

In 1960 as a young technician I was fortunate to be offered a position with an instrument manufacturing company and was given the choice of working 'in pH' or 'in chromatography'.

The choice wasn't too difficult. I didn't know anything about chromatography and –log (hydrogen ion concentration) had no appeal whatsoever. So chromatography it was and I have never regretted the choice.

Like myself in those early days, you may be starting to use gas chromatography for the first time or you may have a little 'hands-on' experience but no real understanding of the theory and practice of the technique. Whichever is the case, the objective of this text is to provide you with a practical appreciation of gas chromatography using the *ACOL: Chromatography Separations* text to provide the theoretical background to the technique. As with any analytical technique there is no substitute for 'hands-on' experience. In gas chromatography the manipulative technique must be learned and practised and the results of experimental changes rigorously studied and understood.

The term chromatography covers those separation techniques in which the separation of compounds is based upon the partition, or distribution, of the analytes between two phases in a dynamic system. In gas chromatography (GC) we have a gaseous mobile phase and a liquid or solid stationary phase.

Where a mobile phase is a liquid, then we are dealing with liquid chromatography (LC)

Π Complete the following table:

Nature of mobile phase	Nature of stationary phase	Name
Gas	Liquid	Gas–liquid chromatography
Gas	?	Gas–solid chromatography
Liquid	Liquid	?
Liquid	Solid	?–solid chromatography

The answers are 'Solid', 'Liquid–liquid chromatography' and 'Liquid–'

The first publication describing gas chromatography was by Martin and James in *Biochem. J.,* **50,** 679 (1952), with the first commercial instruments being manufactured by Griffin and George, Pye, Perkin-Elmer, Wilkins, Hewlett-Packard and others.

Thus gas chromatography as we know it today developed in the late 1950s and early 1960s mainly as a packed column technique although development of capillary columns progressed at the same time. It was not until much later that capillary column gas chromatography became firmly established when fused silica column technology overcame the fragility and variable performance of glass capillary columns.

The introduction of chemically bonded fused silica capillary columns in the 1980s rejuvenated gas chromatography and considerably extended its useful range.

Much of the theory of chromatography was developed in the early years and there was intense rivalry between the leading workers to provide the most convincing arguments.

It was an exciting time for those of us involved.

Nomenclature, definitions and the fundamental parameters of chromatography are to be found in *ACOL: Chromatographic Separations* together with a categorisation of gas, liquid and planar chromatographic systems.

ΙΙ What is the main feature which distinguishes gas chromatography and liquid chromatography from planar chromatography?

Gas and liquid chromatography are both carried out in columns while planar chromatography or thin layer chromatography (TLC) is carried out on plastic, metal or glass sheets coated with silica.

ΙΙ What advantages might planar chromatography, i.e. TLC, have over gas and liquid chromatography?

If you answered 'cost' you would be correct, TLC is cheap, but that was not the answer we were looking for. In gas and liquid chromatography we inject samples into the column but we can never be absolutely certain that all the components are eluted from the column into the detector. In TLC it is possible, by use of visualisation methods, to 'see' all the components on the plate. Nothing need remain undetected.

In this volume we shall be considering gas chromatography, i.e. the system in which the analytes are partitioned between a stationary phase and a gaseous mobile phase. By definition therefore, the compounds to be analysed must be sufficiently volatile for them to be present in the gas phase in the experimental conditions, in order that they may be transported through the column. Fortunately there is generally little association between the vaporised solute molecules and those of the carrier gas which simplifies the chromatographic process.

As we shall discuss later, analyte volatility is one of the major limiting factors in the application of the technique. The basic principle of gas chromatography is that the greater the affinity of the compound for the stationary phase, the more the compound will be retained by the column and the longer it will be before it is eluted and detected.

It is perhaps simplest to remember that all solute molecules spend the same amount of time in the gas phase. The difference in retention times, i.e. how long it takes to elute the compound from the column, depends upon its retention by the stationary phase.

In later sections we shall discuss the nature of the stationary phase, how columns are prepared and how separations may be optimised.

Thus, the heart of the gas chromatograph is the column in which the separation of the components takes place, and to this must be added the source and control of the carrier gas flow through the column, a means of sample introduction and a means of detection of the components as they elute from the end of the column. Since temperature will influence the volatility of the analytes, the column is placed in a thermostatically controlled oven. Detectors and some injectors are also heated. A basic chromatograph is represented diagramatically in Fig. 1a.

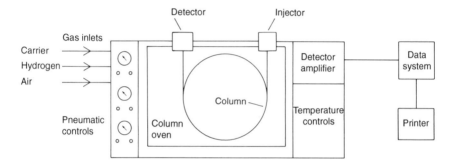

Fig. 1a. *Schematic representation of a basic gas chromatograph*

Since the earliest days of gas chromatography there has been much interest in both the quantitative and qualitative aspects of the technique. It was natural that having been able to separate mixtures of compounds, the questions, 'How much?' and 'What are they?' should soon be asked.

Initially, detectors were very primitive and in many cases the mechanisms of detection were unclear. Some detectors had their origins in gas analysis and their internal geometry was far from ideal for their application in the dynamic, small-volume gas chromatographic environment where concentrations of analyte in the gaseous medium were low and where fast response was essential if the eluting peaks were to be faithfully monitored.

The earliest detectors were based on thermal conductivity, to be followed later by the flame thermocouple, the first flame detector, argon ionisation detector and then the flame ionisation detector.

Others followed, but without question the flame ionisation detector, (FID), has been the workhorse in GC. Detectors will be discussed in Part 6.

How might the structure and elemental composition of the compound influence its response in the detector and how would this affect quantitative analysis? Could peak heights be used as a measure of 'how much' or was it necessary to use peak areas? Would compounds respond linearly with mass injected? Would compounds of similar structure have similar responses? These were some of the questions, sometimes very difficult questions, facing the early gas chromatographers.

Thus, although in truth, quantitative analysis with the early GC system was poor due to non-linear response detectors and rather primitive data systems, things improved as new equipment was developed.

To answer the 'what' question the direct combination of gas chromatography with mass spectrometry (MS) in the early 1960s was a major step forward in compound identification, particularly in the case of complex mixtures. Today GC–MS and GC–infrared (IR) are routine, having been simplified by the technological developments which have taken place.

Π What do you consider to have been the major problem to be overcome in combining gas chromatography with mass spectrometry?

Mass spectrometers work under high vacuum, typically 1×10^{-6} to 1×10^{-8} torr whereas the outlet from the gas chromatographic column is at atmospheric pressure. Interfaces had to be designed to handle these pressure differences, and in the case of packed column gas chromatography where gas volume flow rates were typically 30–50 ml/min, capable of removing the bulk of the carrier gas preferentially, thus increasing the concentration of analyte entering the source in the mass spectrometer.

Those becoming involved for the first time in gas chromatography might conclude that gas chromatography started as a packed-column technique and then in the 1980s when capillary columns were invented, everyone 'went for it' and threw away their packed columns.

Indeed packed-column gas chromatography, as we know it, did become established first in the late 1950s to early 1960s but in fact as early as 1958, Marcel Golay proposed the use of capillary columns.

Development of the capillary column took time but early in the 1960s quite spectacular separations were being obtained for complex hydrocarbon mixtures. Commercial exploitation was hindered for some years by patent problems and, in effect, it was the development of the robust fused silica columns in the 1980s which brought about the resurgence in the application of this technology.

It is appropriate in this book to split the text into sections which deal specifically with aspects of packed and capillary systems.

Although we shall consider the two types of column systems in detail in later sections it is perhaps appropriate to define packed and capillary at this stage.

A packed gas chromatographic column consists of a suitable tube containing an inert solid which has been coated with a relatively involatile liquid phase. The solid acts as a support for this liquid phase, called the stationary phase. A gas, normally inert, i.e. the carrier gas, is passed through the packed bed and differences in the partition coefficients of the individual components in the mixture between the gaseous and stationary phases causes a separation of a mixture of solutes placed at the beginning of the column. The resolving power of packed columns, measured in theoretical plates, may be in the order of a few thousand plates for a column 1 metre long.

In capillary column chromatography, the scale of the chromatographic system is reduced, the column is a narrow capillary, 0.1–0.5 mm internal diameter, 10 m–60 m in length, the solid support is dispensed with and the stationary phase is coated directly onto the column wall.

The chromatographic process remains the same, partition of the solutes between the gas and stationary phase and retention based upon affinity for the stationary phase. Capillary columns frequently generate 50 000–100 000 or even more theoretical plates although it is prudent to bear in mind that resolution is not always simply a matter of theoretical plates. Let me give you an example.

Fig. 1b and Fig. 1c show chromatograms of a simple mixture of fatty acid esters. In the case of Fig. 1b, a 2 m packed column has been used and baseline separation achieved. Fig. 1c shows the same mixture analysed on a capillary column but, although very sharp peaks are

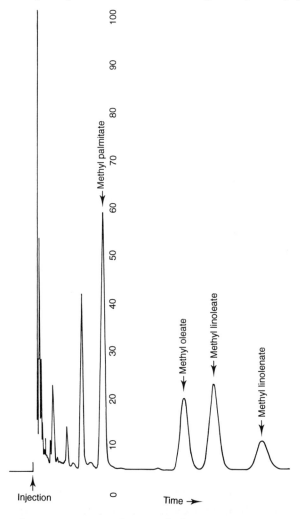

Fig. 1b. *Gas chromatogram of a methyl ester mixture, using a 2 m packed column*

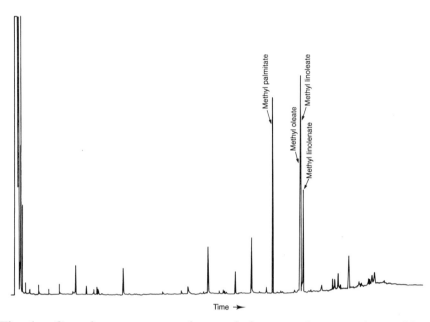

Fig. 1c. *Gas chromatogram of a methyl ester mixture, using a 25 m capillary column*

observed, indicating much greater column efficiency, two components are not resolved. Clearly there is more to separation than just efficiency.

Π Why do you think separation has been achieved on the packed column and not on the capillary?

The packed column contains a more polar stationary phase which produces a greater affinity difference between the closely eluting peak pair compounds than does the capillary column, thus promoting better separation.

The technology of the gas chromatograph reflects the need to optimise the advantages of the different column system, thus whereas injection systems for packed columns are relatively simple, more sophisticated injection devices are required to place the small quantity of sample required by a capillary column into the column, without compromising the potential high separation efficiency of the system.

Modern capillary columns in which the stationary phase is chemically bonded to the silica column wall have extended the temperature range of gas chromatography, effectively expanding the scope of the technique and making possible the analysis of compounds previously considered the preserve of high performance liquid chromatography (HPLC). In addition, polar stationary phases can also be chemically bonded onto the column wall providing an additional degree of selectivity to enhance the resolving power of the large number of theoretical plates.

The advantages of capillary columns are really two-fold. The much greater number of potential theoretical plates means greater resolution of components in complex mixtures. Many organic chemists rue the advent of the capillary column! Chromatographic peaks are much narrower on capillary chromatograms, cf. Figs 1b and 1c, and this enhances sensitivity to detection of low-level components. The combined effect of these two advantages is a higher degree of precision in quantitative analysis.

SAQ 1a List the differences between packed and capillary columns.

SAQ 1a

Without doubt, capillary column gas chromatography requires a much greater degree of operator skill, it is not just a simple case of changing from one column type to another. The skills must be learned by 'hands-on' experience, not just by reading the instructions supplied with the column or by reading this book, no matter how good these may be! Always remember the scale in which you are working and focus your attention to detail accordingly.

Gas chromatography is still a developing technology, particularly in the areas of new stationary phases and in its combination in multi-dimensional chromatographic and spectroscopic systems.

Above all enjoy your chromatography. For me, gas chromatography is still an exciting technique offering new opportunities for solving problems.

Learning Objectives of Part 1

After studying the material in Part 1, you should be able to:

● define gas chromatography and delineate its boundaries;

- discuss the basic principle of gas chromatographic separation;

- describe the structure of a basic gas chromatograph;

- discuss the fundamental difference between packed column and capillary column gas chromatography;

- discuss the relative merits of the two alternative methods;

- realise the potential of gas chromatography for both qualitative and quantitative analysis.

2. The Fundamental Chromatograph

A simple block diagram of a typical gas chromatograph is given in Part 1. In this section we are concerned with a discussion of each component part of the system in turn.

2.1. THE GAS SUPPLY SYSTEM

Gas chromatographs require a supply of carrier gas of sufficient quality and pressure to achieve the desired separations. Carrier gases, usually nitrogen, helium or hydrogen are normally supplied from compressed gas cylinders, although some users of nitrogen obtain their supply from cryogenic liquid nitrogen tanks or from laboratory nitrogen generators designed specifically for chromatography use. Carrier gas should be inert, dry and free of oxygen to prevent degradation of the column.

∏   Could dry compressed air be used as carrier gas?

Air contains approximately 20% oxygen. As discussed above the carrier gas should be free of oxygen.

Line pressures of approximately 30 psi are adequate for both packed and capillary work and this pressure is sufficiently high to allow the gas pressure regulators and flow controllers built into the chromatograph to operate effectively.

For capillary column work, helium is the preferred gas and provided you have ensured that your system is free of leaks, it is not very expensive since a cylinder should last in excess of six months. If you are running your gas chromatograph with a mass spectrometer then you must use helium.

Hydrogen may be used as a carrier gas and it is, in fact, the ideal carrier gas particularly for capillary chromatography. There is a potential explosion risk in using hydrogen although in thirty years I have only heard of one incident which resulted in an explosion. Fortunately, no one was injured. Hydrogen leak detectors can be installed but unfortunately they are not totally specific and do give rise to false alarms. Oxygen and moisture traps can be incorporated into the carrier gas line as an additional precaution.

SAQ 2.1a	Which gases are normally used for gas chromatography?

If you are using a flame ionisation detector, the hydrogen and air supplies must also be arranged either from cylinders, generators or pumps.

Π Why should only high-purity gases be used?

Apart from the problems with oxygen and moisture in the carrier gas, it is important that gas supplies should be free of other contaminants which might be detected by the detector. Organic impurities in the carrier or in the hydrogen and air to the flame ionisation detector may give rise to a significant base current response which may reduce the sensitivity of the detector.

SAQ 2.1b

> Could the following gases be used for gas chromatography and if so do they have any particular limitations or advantages; hydrogen, argon and carbon dioxide?

Do spend some time checking your gas line for leaks. Analytical chemistry costs companies and organisations a lot of time and money. We are the custodians of the resources entrusted to us, so do make every effort to eliminate waste, be it time, money or gas.

2.2. THE COLUMN OVEN

This is normally an accurately temperature controlled fan-blown oven with adequate space to accommodate columns and with an even temperature distribution throughout the oven. Ovens can normally be temperature programmed at a variety of rates with isothermal periods set as desired.

Larger ovens are generally easier to work in, particularly when installing columns.

∏ Suggest two additional desirable characteristics of the chromatograph oven.

(1) Rapid temperature response to follow accurately the temperature programme profile and (2) low thermal mass for fast cool-down at the conclusion of the analysis. On extended temperature programme cycles this can influence sample through-put.

2.3. THE INJECTION SYSTEM

The injection system provides a means of putting the sample or the sample solution onto the column. With packed columns this is normally a simple septum injector, but for capillary column gas chromatography, i.e. high resolution gas chromatography (HRGC), a number of different injection systems are available. Frequently, new capillary instruments are fitted with two different injectors.

∏ In addition to allowing us to inject our sample into the chromatographic column, what is another function of the injector?

The injection device provides a means of maintaining the pressure and flow through the analytical column during the injection process and at the same time prevents the ingress of air into the column system.

The principle capillary injection devices are the 'split/splitless' injector and the 'on-column' injector.

A sample injected into the splitless injector is vaporised in a heated chamber and a small and (hopefully) representative portion allowed to pass into the capillary column. Using the on-column system, the sample is injected into the column using a syringe fitted with a very thin needle through a suitably engineered injection port. The objective of injection methods is to place a narrow band of solutes at the beginning of the column in order to maximise the resolving power of the column. Capillary injectors will be described in Part 5.

SAQ 2.3a

> Why must different injection systems be used in capillary column gas chromatography, i.e. HRGC?

Some injectors are fitted with heaters which are quite separate from the column oven. Use injector heaters sparingly, it is not normally necessary to set them at a temperature which exceeds the maximum analysis temperature by more than 5 degrees. It is worth experimenting with injection temperature, as a reduction of 5 or 10 degrees may prevent some sample decomposition, improve quantitative analysis and prolong the intervals between column maintenance.

There have been many debates and arguments about injectors, injection heaters and temperatures, etc. Sample introduction is a critical part of the chromatographic process and is one of the skills the analytical chemist must develop.

2.4. THE CHROMATOGRAPHIC COLUMN

Columns will be either packed or capillary and these will be described in detail in Part 3 and Part 4, respectively.

It is important that the column system should not be stressed in the oven and that couplings should not be over tightened. It is also good practice to check them again after a few hours use.

Ensure you are aware of your column's temperature limitations and if it is possible, programme in an oven cut-off temperature in case some other operator should unwittingly try to exceed the prescribed limit.

∏   Where would you find the temperature limit of your particular column?

Chromatography supply catalogues are the most convenient source of stationary phase and column temperature range data. New columns are normally supplied with a test certificate which will state the recommended operating temperature range. Some manufacturers also provide useful hints and information on how to get the best out of your column.

Check the column flow rate using a simple soap bubble flow-meter prior to connecting the column to the detector. Note the pressure setting which gives you the desired flow. Capillary column flows can also be checked in this way using a suitable small volume flow-meter. When you have more experience working with capillary columns, you will place more emphasis on the inlet pressure and simply check that the flow is in the right order by immersing the column end in a suitable volatile solvent and observing the stream of bubbles.

Having installed the column, it is important to carry out a leak test on the complete system systematically. Avoid using soap solution, since if even a very small quantity should enter the gas flow line then this may give rise to baseline and column problems. Ethyl acetate (ethyl ethanoate) is a suitable solvent since it is completely volatile. Check each connection in turn and, should bubbles appear, then tighten the connection slightly. If connections have been correctly assembled then excessive force is not necessary.

An excellent alternative method of leak testing is to block the detector outlet with a suitable fitting, pressurise the complete system, then isolate the inlet supply. If the line pressure remains constant then the system is gas tight, however if the pressure drops it will do so at a rate commensurate with the magnitude of the leak. We used to call this 'The Gas Works Test'.

2.5. DETECTORS

The detector is the measuring device in the chromatographic system, and as the name implies, it detects the presence of compounds in the gas stream as it leaves the column.

Detectors are located in a separately controlled heated zone in the instrument. Flame ionisation detectors (FIDs) are the most common ones in use today as they are robust and sensitive and generally give rise to few problems.

The range of detectors which may be used in gas chromatography is discussed in Part 6.

2.6. AMPLIFIERS

The signals or responses generated by gas chromatography detectors are very small and must be enhanced electronically to make them 'visible' on the recorder or data system. This is the function of the detector amplifier.

The operation of the amplifier need not concern us here. It is important to ensure that it is connected as instructed by the manufacturer, if it is a stand-alone unit, but in many instruments today it is built in and is an integral part of the system. The operator should be familiar with the significance of the range and attenuation controls and operate the unit in such a way that the data system receives the optimum signals. Some data systems prefer a 1 volt unattenuated input while others are happy with 10 millivolts. Check your manuals and then test it out.

2.7. DATA SYSTEMS

It would be impossible to deal with the topic of data systems comprehensively in this book since the range and complexity of these systems is enormous.

Data systems normally have a number of calculation routines pre-programmed which reduces the need for manual calculation and data tabulation.

It is fundamental that your system handles and manipulates the signal it receives precisely and accurately. It is not within your power to check, or validate, the data systems operation as a stand-alone, irrespective of what some pharmaceutical quality assurance people would have us believe. These are electronic data handling systems and can only be validated with the appropriate electronic equipment and software. The fact that they are on the market is justification of their suitability. If you obtain results which appear not to reflect the chromatographic information as you see it, then it is probable that your parameters have been entered incorrectly. It is prudent to retain a print-out of your integration parameters and file it with your chromatograms and data.

∏ Your chromatogram appears as a simple trace on a potentiometric recorder. How could you generate some quantitative data very simply?

Peak heights are an indication of the relative amounts of components in the mixture. Peak areas can be measured by triangulation or by cutting and weighing and are approximately proportional to the mass of the component passing through the detector.

Data handling and quantitative analysis will be discussed in Part 7.

2.8. STARTING UP THE COMPLETE SYSTEM

All the components discussed form the working chromatographic system. When starting up the instrument, it is good to have a routine, you may even prepare a checklist. The list below assumes that you have already checked for leaks and that your column has been installed correctly.

Checklist

(1) Turn on the gas supplies and set the carrier gas pressure to give the desired column flow rate.

(2) Switch on the mains power to the instrument and associated units. Ensure that all expected power lights are illuminated.

(3) Set the column oven temperature and also the temperature programme conditions if appropriate.

(4) Set the injector temperature.

(5) Set the detector temperature.

(6) Allow time for the injector, the detector and the column oven to reach the set temperature.

(7) If an FID is fitted, set the appropriate hydrogen and air pressure.

(8) Ignite the detector and check.

(9) Set amplifier range and attenuation as required.

(10) Set the required integration and presentation parameters into the data system and check.

(11) Put the data system into run and zero the amplifier as prescribed in the operating manual.

(12) Inject a sample of solvent and monitor the response on the recorder or the data system.

If the solvent response is good then your system should be ready for use.

When checking for faults be systematic and remember, if all else fails, read the instruction manual!

SAQ 2.8a   List six items which should be checked when starting up a gas chromatograph.

Learning Objectives of Part 2

After studying the material in Part 2, you should be able to:

- discuss the role of each of the component parts of the working system;

- organise the services necessary to set up a working chromatograph;

- start up and check the satisfactory operation of a gas chromatograph.

3. Packed Column Systems

3.1. INTRODUCTION

The chromatographic column is the most important part of any gas chromatograph.

Many years ago I was a member of a team assigned to designing a chromatographic column to fit into an ideal column oven which had been designed by the company's engineering department.

The project was flawed; the engineers knew how to design a good oven in terms of temperature profiles etc., but they knew little about chromatography. We knew a lot about chromatographic columns but were unable to design a suitable configuration, to fit in the ideal oven, which matched our expectations for column performance in a new instrument.

Needless to say the instrument was a failure, primarily because those managing the project were not good chromatographers!

It is important to remember to consider all the component parts of the gas chromatograph as ancillaries to the column; if we do not get the column correct, the rest are not going to help.

Gas chromatography started with packed columns although as you will realise when you progress to Part 4, capillary columns are now more generally used. In the following section we will consider the dimensions of columns, the nature of the packing material, the mechanism of separation and the optimisation of conditions and separations.

3.2. COLUMN DIMENSIONS

Most packed columns are glass, 1.5 m–10 m long and 2 mm–4 mm internal diameter. Glass is easily manipulated to specific designs, is robust and being transparent enables us to ensure that the column when packed is complete, without gaps or channels. In one laboratory in which I worked we built 60 ft glass columns for flavour analysis and these separated complex mixtures very successfully once we had designed a suitable high-pressure injection head.

Metal columns are frequently used for industrial plant monitoring chromatographs where safety considerations and robustness are of paramount importance.

∏ How might the risk of gaps in the column packing in a metal column be reduced?

The optimum packing density (g/ml) can be determined for the packing in an equivalent internal diameter glass column. The internal volume of the metal column can be calculated ($\pi r^2 h$) and hence the weight of packing which must be used to completely fill the column, determined. It only remains then to pack the column with that weight of material.

Early gas chromatographs had straight columns which were very easy to pack and gave good efficiencies. As designs for bench-top instruments became popular, coiled columns were developed but these never gave as high efficiencies as their straight counterparts.

The loss of efficiency in coiled columns compared with straight is a function of packing density and not as initially feared that the molecules travelling on the inside of the bend would reach the end of the column before those on the outside, resulting in peak broadening. This effect is negligible provided that the coil diameter is at least ten times the internal diameter of the column tubing.

Columns must terminate in a suitable fitting to allow coupling to the detector with minimal dead volume. Glass-to-metal seals are generally used although in some instruments the end of the column connects directly to the detector base. It is important that there is a frit or plug of silanised glass or quartz wool at the end of the column to prevent packing being blown into the detector.

SAQ 3.2a

> Glass and stainless steel tubing may be used in the manufacture of packed columns for gas chromatography. List two advantages and one disadvantage for each.

Injection systems for packed column gas chromatography consist of a gas inlet and a septum injection port (see Fig. 3.10a).

3.3. COLUMN PACKING

Two specific types of packing are used in gas chromatography.

An inert solid support coated with a film of non-volatile liquid which is the active stationary phase. This is gas–liquid chromatography (GLC).

The second type of packing is an uncoated solid, which may be a simple adsorbent or a microporous solid such as molecular sieve. This would be gas–solid chromatography (GSC).

The particle size and size distribution of the packing material influences the performance of the column and the inlet pressure necessary to obtain optimum flow rates.

Particle sizes are presented in mesh size ranges:

Particle diameter	Mesh size
0.1–0.125 mm	120–140 mesh
0.125–0.15 mm	100–120 mesh
0.15–0.25 mm	60–100 mesh

Remember that the smaller the particle size used in your column, the more difficult it becomes to drive the carrier gas through. The size distribution should be as narrow as possible, and during preparation of the packing, care should be taken to avoid breaking the particles since fines will result in an inhomogeneous packed bed.

Π   What other problems might be caused when particles break off the column packing material?

Fines may restrict the gas flow and even 'blow' into the detector and block the narrow passages or the jet in an FID. In addition, the disintegration of coated particles is likely to expose active sites on the solid support and these may give rise to solute adsorption which could show up as tailing peaks.

3.4. SOLID SUPPORTS

The solid support acts as a medium for distributing the stationary phase such that it is exposed to the carrier gas and thus to the solute molecules in the gas phase. Although the solid support in gas–liquid chromatography takes no part in the chromatographic process, the following properties are necessary for an ideal support.

- Relatively large surface area per unit volume.

- Chemically inert, thermally stable and low-to-zero adsorption properties.

- Mechanically robust, in order to withstand coating and packing procedures without breaking down to expose an uncoated surface.

- Uniform, spherical (if possible) particles with a narrow size distribution range.

- A pore structure favourable for promoting fast mass transfer (no long tunnels!).

The common GC supports are based on diatomaceous earth, the fossilised remains of single-celled plants. These skeletal residues have a highly porous structure which makes them ideal as support material.

Chromosorb W is a white solid prepared by heating the diatomaceous earth with a flux which results in much of the fine structure being cemented over by a glass, resulting in large and irregular pores. This support is also relatively fragile and must be handled carefully.

Chromosorb P is made by crushing the diatomaceous earth, pressing it into bricks and calcining at about 900 °C. The resulting pink brick, better known as Johns-Manville Firebrick, is then ground up and sieved for use as solid support. As a lad, I remember spending many hours breaking up these bricks and then 'mincing' them through a

Zacharia-Parkes Coal and Coke Grinder. After washing and sieving, everything in the lab was covered in a fine layer of pink dust. I also recall that we produced a superior grade of packing by heating our solid support to 1200 °C in a muffle furnace and we demonstrated by electron microscopy that the internal structure was radically altered. Chromosorb P is a more robust support than chromosorb W and can hold up to 30% stationary phase.

Solid supports can be supplied acid washed and silanised. Silanisation deactivates the support by removal of the active hydroxyl groups.

$$
\begin{array}{c}
\text{O} \\
\diagdown \\
\text{Si-OH} \\
\diagup \\
\text{O} \\
\diagdown \\
\text{Si-OH} \\
\diagup \\
\text{O}
\end{array}
\quad + \quad
\begin{array}{c}
\text{Cl} \diagdown \quad \diagup \text{CH}_3 \\
\text{Si} \\
\text{Cl} \diagup \quad \diagdown \text{CH}_3
\end{array}
\quad \longrightarrow \quad
\begin{array}{c}
\text{O} \\
\diagdown \\
\text{Si-O} \\
\diagup \quad \diagdown \quad \diagup \text{CH}_3 \\
\text{O} \qquad \text{Si} \\
\diagdown \quad \diagup \quad \diagdown \text{CH}_3 \\
\text{Si-O} \\
\diagup \\
\text{O}
\end{array}
\quad + \; 2\,\text{HCl}
$$

Π Could carbon particles be used as a solid support for gas chromatography?

Carbon is a good adsorbent and can be used in gas–solid chromatography but it could not be classified as an inert solid support for a liquid stationary phase.

There are a considerable range of supports on the market but for general use I would recommend Chromosorb W-HP with a particle size range of 100–120 mesh.

SAQ 3.4a

> List three respects in which diatomaceous earth falls short of the ideal as a support for gas chromatography.

SAQ 3.4a

3.5. PREPARATION OF COLUMN PACKING

Coating of the stationary phase onto the solid support can be carried out successfully in the laboratory at very little cost although most chromatographers prefer to buy ready coated packing from a reputable supplier.

If you decide to try it yourself there are two approaches.

I have always preferred to use an open beaker, weigh in the solid support and the stationary phase and then add sufficient organic solvent to obtain a mobile slurry. Leave the slurry for about thirty minutes with occasional swirling of the mixture to allow the stationary sufficient time to dissolve and disperse throughout the solid support. To evaporate the solvent hold the beaker in a hot water bath, in a fume cupboard, and keep the contents moving gently. Do not stir with a spatula. The idea is to remove the solvent without breaking up the particles. Occasionally you may need to ease the material adhering to the walls, but finally you will have a free flowing powder free from lumps and aggregates and ready for final drying in an oven. Do not worry if it does not work first time, you can always add more solvent and start again.

The alternative is to put your slurry in a rotary evaporator and remove the solvent by heat and vacuum while the flask is rotating. This is a bit aggressive, but is obviously a practical proposition in a commercial environment where large quantities of packing are being prepared. Packing prepared this way is best gently sieved to remove any fines before packing.

Π   I require column packing with a 10% loading of stationary phase. I have used:

   10 g solid support + 1 g stationary phase.

   Am I correct?

No! This will give me 9.1% on a wt/wt basis. This is a very common mistake.

The appropriate solvents for dissolving the stationary phases are listed in the suppliers' catalogues.

Packing the column requires time and some patience if you are to get the maximum efficiency from your end-product.

Make sure that the column is clean and dry. If necessary, wash it though several times with solvent, say dichloromethane and acetone and blow dry with nitrogen. Finally dry in an air oven at 110 °C. Remember the plug at the detector end. Fit a small funnel to the inlet end and connect the detector end to a vacuum source.

Add packing, a few grams at a time, and tap the column gently with a wooden ruler or a balance brush. The packing will move through the volume to the detector end and compact into a bed. Continue adding packing and tapping until the column is full and the bed no longer packs down.

Ensure that there are no gaps.

A plug of silanised glass wool at the injection end of the column is sometimes used to keep the packing in place.

Turn off the vacuum source and allow the pressure in the column to equilibrate before disconnecting. Otherwise your packing will end up all round the lab! Vapour phase chromatography?!

Now mount your column in the column oven but do not connect the detector end. Establish flow at say 40 ml/min and again tap the column gently. You will probably find that the packing settles down further in which case it will be necessary to stop the flow, remove the column and top up.

∏ Why do you think the column should not be connected to the detector at this stage?

During the packing process small particles may pass the end plug and could block the detector.

The next step is to condition the column, i.e. preheat it with carrier gas flowing.

With carrier gas flow set, but with the column still not attached to the detector, set the column oven to temperature programme slowly to the maximum operating temperature for the phase. If you do not need to go to the maximum temperature for the analysis you intend carrying out, then programme to approximately 10 degrees above the final temperature. It is better to run-in a column gently if you can and gradually work up to the temperature maximum over a period. After, say, a couple of hours at your top temperature, open the oven and examine the column carefully. Look for gaps and again tap the column gently and if necessary add more packing. Most texts recommend conditioning columns overnight but you will probably find performance to be satisfactory after a few hours, enabling you to get your analysis started with final conditioning taking place overnight.

∏ Why should the column not be connected to the detector during the conditioning period?

Conditioning the column involves purging out solvent and stationary phase material which might interfere with the later analysis. Some of this material could possibly condense in the detector or deposit as residues on the flame jet of an FID detector and might subsequently give rise to detector 'noise'.

The end of your packed bed should be about 5 mm above the point of discharge of the syringe needle. I am going to recommend that you inject into the packing, but more of that later.

Your column should now be ready for use.

3.6. COLUMN PERFORMANCE

In Section 3.5 we have described the preparation of column packing and the packing and conditioning of the column. Now comes the time to assess whether the column is going to be satisfactory for the job.

In general it is good practice to measure three parameters using an appropriate sample mixture chromatogram run isothermally at as near as possible to the optimum column flow rate.

(1) Column efficiency, either measured as the number of theoretical plates *(N)*, sometimes referred to as the plate count or the plate number, or the plate height *(H)*.

(2) Measure the resolution factor (R_S) for two closely eluting peaks if you have them in your mixture.

(3) Measure peak symmetry.

If you are unsure of the meaning of these terms then refer to *ACOL: Chromatographic Separations*, Part 3.

Column efficiency may be measured from the peak profile in a number of different ways, the most common using the construction and measurement shown in Fig. 3.6a and the following relationships:

(a)
$$N = 4 \left(\frac{t_R}{W_{0.6065\,h}} \right)^2$$

(b)
$$N = 5.54 \left(\frac{t_R}{W_{0.5\,h}} \right)^2$$

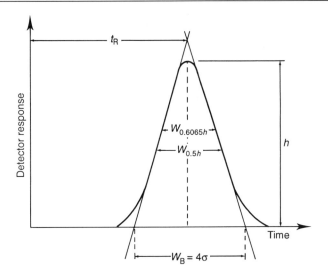

Fig. 3.6a. *Measurement of N*

(c)
$$N = 16 \left(\frac{t_R}{W_B} \right)^2$$

Where

t_R = uncorrected retention time

$W_{0.6065\,h}$ = peak-width at the point of inflection

$W_{0.5\,h}$ = peak-width at half peak-height

W_B = peak-width at the peak base

Peaks must be on-scale in order to make these measurements. This may seem obvious but I have seen it attempted with peaks well off-scale!

Alternatively using area and peak height counts from a digital integrator print-out then:

$$N = 2\pi \left(\frac{t_R \times \text{peak height counts}}{\text{peak area counts}} \right)^2$$

These calculations overestimate the column efficiency and if preferred the corrected retention time may be used, in which case the result becomes the *effective plate number* (N_{eff}).

It should be possible to obtain efficiencies of approximately 3000 theoretical plates for a 1.5 m × 4 mm internal diameter (i.d.) column. Generally, early and late peaks in the chromatogram will show lower efficiency, the maximum usually occurring at 8–12 minutes into the run.

The plate height H is the *height equivalent to one theoretical plate,* hence *HETP.* In fact, it is the length of column equivalent to one theoretical plate since it is calculated by dividing the length of the column *(L)* by the number of theoretical plates *(N)*:

$$H = L/N$$

The plate height measurement is normally used in the construction of *HETP* curves where optimum gas velocity is being determined for a particular column.

Resolution factor (R_S) is a measure of the degree of separation of adjacent peaks and is determined from measurements taken from the peak pair as shown in Fig. 3.6b and by using the calculation below.

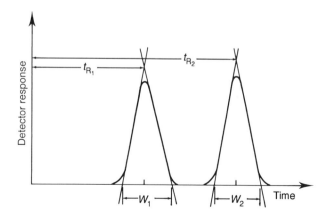

Fig. 3.6b. *Measurement of resolution*

$$R_s = \frac{t_{R_2} - t_{R_1}}{\frac{1}{2}(W_1 + W_2)}$$

where

t_{R_1} and t_{R_2} are the peak retention times

W_1 and W_2 are the widths at the peak bases.

Baseline separation of symmetric peaks is considered to have been obtained if the value of R_s is 1.5 or less.

The peak asymmetry factor (A_S) is a measure of peak distortion, sometimes referred to as the skew factor. Measurements are taken as shown in Fig. 3.6c and the factor calculated as follows:

$$A_S = \frac{CB}{AC}$$

Ideally, values of very nearly 1.0 should be obtained. If the value is 1.2 then this indicates poor column packing. If it is greater than 1.6, discard the packing and start again. If the asymmetry factor is less than 1.0 then this will indicate that the peak may be showing signs of overloading. This can be easily checked by reducing the injection volume and running the chromatogram again. If the value of A_S is now 1.0 or greater, then overloading has been eliminated.

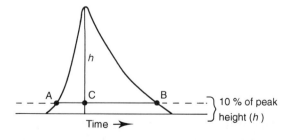

Fig. 3.6c. *Measurement of peak asymmetry*

∏ Now try using these relationships to calculate the efficiencies of the methyl oleate and methyl linoleate peaks on the chromatogram shown in Fig. 1b. Also calculate the resolution factor for this peak pair.

You may find it useful to enlarge the chromatogram on a photocopier before you draw your constructions and make measurements.

Calculating the efficiency at half-peak height I obtained the following:

methyl oleate   1500 theoretical plates.
methyl linoleate   1800 theoretical plates.

I calculated the resolution factor to be 1.83.

You could also try calculating the peak asymmetry of the peaks but you will find that they both come out very close to 1.0

3.7.   THE STATIONARY PHASE

It may seem odd that I discussed how to prepare the column packing without first considering the nature and role of the stationary phase. This was deliberate since the method of preparation is, to all intents and purposes, independent of the stationary phase character apart from the choice of solvent.

In this section we shall consider the range of stationary phase types and their role in the chromatographic separation.

In many respects, your study of this section of the book will provide the basis of your understanding of the mechanism of the gas chromatographic process. You should, in addition, reinforce your knowledge by studying the relevant sections of *ACOL: Chromatographic Separations*.

Although column efficiency, i.e. the number of theoretical plates in the column, plays an important role in the separation achievable by the column, it should be remembered that it is selectivity which determines how well the stationary phase differentiates between substances.

Selectivity is defined as

$$\alpha = K_a/K_b$$

where K_a is the distribution constant of substance A
K_b is the distribution constant of substance B

In making our choice of stationary phase we are attempting to influence α such that our compounds are more likely to be resolved. Later as we get into the separation, we can further influence the selectivity by choice of column temperature.

The use of liquid stationary phases in gas chromatography has a number of advantages.

(a) Liquid phases are available in great variety thus allowing suitable selectivity for a particular separation.

(b) Liquid stationary phases are available in well defined purity enabling reproducible retention times to be produced.

(c) The amount of stationary phase on the column can be varied to suit either analytical or preparative separations.

All conventional stationary phases have one major disadvantage, their volatility. Thus, if they are used above their recommended upper temperature limit a certain amount of liquid phase will bleed into the detector causing baseline drift, loss of sensitivity and a change in the nature of the column.

Manufacturers frequently report both maximum and minimum operating temperatures. The minimum temperature is important since below this temperature the stationary phase becomes a solid and as a solid it may have significantly different characteristics from those of the liquid state. It is not unknown for a separation to be lost altogether when the stationary phase becomes solid, with the compounds simply not being retained on the column.

Stationary phases are normally materials of relatively high molecular weight, such as saturated hydrocarbons, silicones, ethers, esters and amides.

In order for any retention of the solute molecules to take place, it is necessary for the solute molecule to dissolve in the stationary phase for a finite period of time before re-emerging from the liquid surface into the vapour state.

Solubility of solutes in stationary phases can therefore become a starting point in our process to choose a suitable stationary phase for a particular solute mixture.

Organic chemists are masters of crystallisation and they will tell us that *like dissolves like*. For example, saturated hydrocarbons are most soluble in other saturated hydrocarbons but not very soluble in alcohols, etc., while alcohols are soluble in other alcohols but are not very soluble in hydrocarbons.

On this principle we would predict that the solubility of the components of a mixture in the stationary phase would be proportional to the similarity between their chemical composition and that of the stationary phase. Thus compounds which are similar in nature to the stationary phase will be retained in the liquid phase while those chemically dissimilar will pass more rapidly through the column.

SAQ 3.7a	By applying the principle of *like dissolves like*, for each of the following pairs of compounds, suggest which one will be most soluble in the stated solvent. (a) naphthalene and 1-naphthol in benzene (b) benzophenone and diphenylmethane in propanone (c) methanol and butan-1-ol in water.

SAQ 3.7a

With the *like dissolves like* approach there are pitfalls for the unwary. Were we to consider a mixture of 2-nitroanaline with 4-nitroaniline we might well have considered that both were nitroamines and we could therefore expect them to have similar solubilities. In fact if we test these compounds for solubility we find that 4-nitroanaline is much less soluble than 2-nitroaniline. Clearly some other phenomenon is at work.

Our organic chemist, busily choosing solvents for recrystallisation, is more likely to think in terms of the *polarity* of the solvents and the solutes than to use the simple *like dissolves like* approach. He will assume that a polar solute will dissolve best in a polar solvent and vice versa. What he understands by *polarity* is a somewhat loosely defined concept which combines together the dipole moment of a compound and its hydrogen bonding ability. Thus hydrocarbons with no dipole moment and no hydrogen bonding ability, are held to be non-polar. Water, which has a reasonably large dipole moment and a very strong hydrogen bonding ability, would be reckoned to be very polar, in fact it is one of the most polar solvents we have. Alcohols, with similar dipole moments and rather less hydrogen bonding ability, because they only have one hydrogen atom attached to the oxygen, are less polar than water. Ketones and esters may have higher dipole moments than alcohols because of their carbonyl groups, but they are not strongly hydrogen bonding. They can form intermolecular hydrogen bonds only to other molecules which can supply the necessary hydrogen atom (i.e. act as hydrogen donors):

$$
R^{\diagup O \diagdown}H\text{--}X_{\diagdown H} \quad \text{and} \quad \text{but} \quad R^{-C-R^1}_{\underset{\displaystyle O}{\overset{\displaystyle H-X}{\|}}}
$$

$$
\begin{array}{c} R \\ \diagdown \\ O\text{--}H \\ \diagup \\ H \end{array} \quad X \diagup
$$

Ketones and esters are therefore less polar than alcohols. Ethers are less polar again, having a lower dipole moment than ketones because they have no carbonyl group and weaker hydrogen bonding, due to the lower electron density on the oxygen atom:

$$
R^{\diagup O \diagdown}_{R^1}\mathord{\cdot\cdot}H-X
$$

Halogenated hydrocarbons fit between the ethers and the hydrocarbons. They may, like dichloromethane, have a very high dipole moment, but without any significant hydrogen bonding they are not considered to be very polar.

SAQ 3.7b

By applying the principle of *like dissolves like* for each of the following pairs of compounds, suggest which one will elute first from a column of the stated stationary phase.

(a) methylbenzene (bp = 100 °C) and ethyl-2-methylpropanoate (bp = 110 °C) on squalane (a saturated hydrocarbon)

(b) butan-1-ol (bp = 116 °C) and 4-methylpentan-2-one (bp = 117 °C) on glycerol

(c) hexane (bp = 68 °C) and 1-methylethyl methanoate (bp = 68 °C) on PEG-S (polyethylene glycol succinate).

SAQ 3.7b

The cases of 1,2-nitroaniline and 1,4-nitroaniline and their differing solubilities discussed earlier are a special case of hydrogen bonding.

1,4-Nitroaniline is polar and will form hydrogen bonds with the solvent or with other 1,4-nitroaniline molecules. Where hydrogen bonding takes place with the solvent the nitroaniline dissolves, but if intermolecular bonding between 1,4-nitroaniline molecules has taken place, solubility will be inhibited.

In the case of 1,2-nitroaniline, the hydrogen bonding requirement of the molecule can be satisfied by intramolecular hydrogen bonding. Intramolecular bonding forces are greater than intermolecular forces thus there is little tendency for 1,2-nitroaniline to form intermolecular bonds to other nitroaniline molecules and thus there is no requirement for hydrogen bonding with the solvent to achieve solution.

Π   Rearrange the following list of solvents into an order of increasing polarity.

water, iodomethane, ethyl hexanoate, octan-1-ol, hexane, ethanol, propanone.

The correct order, beginning with the lowest polarity, is:

hexane, iodomethane, ethyl hexanoate, propanone, octan-1-ol, ethanol, water.

See if you agree with this reasoning. Hexane, as the only alkane, is totally non-polar, having no dipole moment and no hydrogen bonding ability. Iodomethane has a dipole moment by virtue of the inductive effect of the halogen, but still has little or no hydrogen bonding ability, and so is slightly polar. Ethyl hexanoate and propanone are both more polar because of the combination of the dipole moment and proton acceptor hydrogen bonding ability of the carbonyl groups. Ethyl hexanoate is the less polar of the two, the effect of the polar carbonyl group being diluted by the large non-polar alkyl chain. Octan-1-ol and ethanol, the alcohols, are more polar because of the dipole moment and both the proton donor and the proton acceptor hydrogen bonding ability of the hydroxyl group. Again, octan-1-ol, because of its larger alkyl group, is the less polar of the two. Water, as the best hydrogen bonding compound of all, is the most polar.

What we have done is to arrange the solvents as a *eluotropic series*. This is a series of solvents arranged in increasing order of their ability to elute a component from an adsorbent in a chromatographic column, an ability which is directly proportional to polarity. Such a series is very useful when it comes to choosing a solvent to use for elution in liquid chromatography. If a component elutes too rapidly with one solvent, then you move back in the series; if it elutes too slowly, you move up the series.

If we could arrange a similar classification by polarity of the stationary phases used in gas chromatography, we might find it equally useful in predicting the chromatographic behaviour of components and choosing the right column for an analysis.

Such a classification would range from the non-polar hydrocarbons through the intermediate silicone oils and greases to the slightly polar phthalate esters of long-chain alcohols, then on to the more polar polyethers, ending with the very polar polyesters.

Indeed, if you consult most stationary phase supplies catalogues, this is the way in which the phases are normally listed. Table 3.7a gives a selection of stationary phases in general use.

SAQ 3.7c

> Place the following stationary phases in order of increasing polarity:
>
> polyethylene glycol succinate (PEG-S),
>
> polyethylene glycol — relative molecular mass 400 (PEG 400),
>
> polyethylene glycol — relative molecular mass 20 000 (PEG 20M),
>
> hexadecane,
>
> tritolyl phosphate,
>
> polypropylene glycol adipate (PPGA).

Table 3.7a.　*Stationary phases for gas–liquid chromatography*

Stationary phase	Polarity	Upper limit (°C)	Solvent	Comment
Squalane (2,6,10,15,19,23-hexa-methyltetracosane)	NP	150	hexane	
Apiezon L (high-vacuum stopcock grease)	NP	250–300	dichloromethane	
Silicone oils and gums				
polymethyl siloxanes				
OV 1	IP	350	trichloromethane	
DC 200	IP	200	methylbenzene	
SE-30	IP	300–500	methylbenzene	
polymethylphenyl siloxanes				
OV 17	IP	350	propanone	
SE-52	IP	300	methylbenzene	
polyfluoropropyl siloxanes				
OF1	SP	240	methylbenzene	
polycyanopropyl siloxanes				
OV 105	P	250	propanone	
polycyanopropylmethylphenylmethyl siloxanes				
OV 225	P	250	propanone	
Dinonyl phthalate (DNP)	SP	150	propanone	other phthalate esters also used
Polyethylene glycols				
PEG 400	P	100	methanol	$RMM = 400$; high proportions of terminal OH groups
Carbowax 20M	P	200	methanol	$RMM = 20\,000$; fewer terminal groups
Polyesters				
polyethylene glycol succinate (PEG-S)	VP	180	trichloromethane	
polydiethylene glycol succinate (PDEG-S)	VP	190	trichloromethane	adipates etc. also used

NP = non-polar; IP = intermediate polarity; SP = slightly polar; P = polar; VP = very polar; RMM = relative molecular mass

SAQ 3.7c (*cont.*)

Now let us see how we might use this concept.

If we consider the separation of hexane and 1-methylethyl methanoate, both with boiling points of 68 °C, we can see that we are attempting to separate a non-polar compound from a moderately polar compound. If we used an intermediate polarity or a slightly polar phase column we would probably not resolve the components since their volatilities and their solubilities are essentially similar. In order to achieve separation we must use either a non-polar or very polar stationary phase in order to differentiate the characteristics of the compounds. Thus, if we use a non-polar phase, hexane will be readily soluble and be retained while the ester will be much less soluble and so pass through the column quickly. In the situation where we employ a very polar column, the ester will dissolve and be retained but hexane will be much less soluble and will therefore exhibit limited retention and pass through the column well before the ester.

The net result will be that the elution order will be reversed, Fig. 3.7a(i).

Were we to add methyl ethanoate (bp = 57 °C) to the mixture then while separating the esters on non-polar, intermediate polarity or very polar columns is not a problem, only in the case of the non-polar phase is the solubility of hexane sufficient to ensure an adequate retention which will not interfere with the ester peaks, Fig. 3.7a(ii).

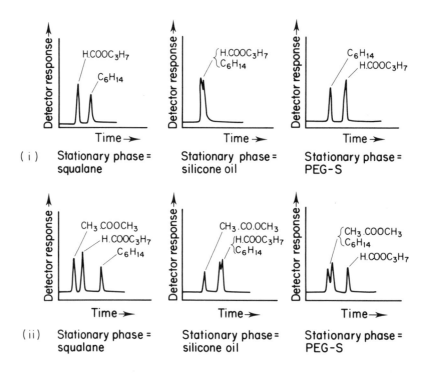

Fig. 3.7a. *Effect of stationary phase polarity on separation of (i) an alkane and an ester of similar volatility; (ii) an alkane and two esters of different volatility*

SAQ 3.7d

Benzene, cyclohexane and ethanol are to be separated by GLC. Given that they all boil between 70 °C and 80 °C, indicate by circling either T for true or F for false whether you agree with either of the following statements.

(1) On a squalane stationary phase at 70 °C the order of elution would be:

ethanol, followed by benzene, followed by cyclohexane.

T / F

(2) On a polyethylene glycol succinate (PEG-S) stationary phase at 70 °C, ethanol would elute after both benzene and cyclohexane.

T / F

You might like to consider what would happen if PEG 400 was used as a stationary phase instead.

Unfortunately some polar compounds exhibit *tailing* when chromatographed on columns packed with non-polar stationary phase, making accurate quantitative analysis more difficult. There is really no simple solution to this problem and therefore wherever possible it is better to separate polar compound mixtures on polar columns, however, there may be circumstances where tailing peaks which are separated are preferable to no separation at all.

SAQ 3.7e

For the three mixtures below, select in each case the answer (a), (b) or (c) which you think is the correct order in which the named components of the mixture will elute on the given stationary phase.

(1) Cyclohexane (bp = 81 °C) and cyclohexene (bp = 83 °C) on dinonyl phthalate (DNP).

    (a) More or less together.
    (b) Cyclohexane then cyclohexene.
    (c) Cyclohexene then cyclohexane.

(2) Methoxybenzene (anisole, bp = 154 °C) and 1-methylethylbenzene (cumene, bp = 152 °C) on polyethylene glycol (PEG 400).

    (a) More or less together.
    (b) Anisole then cumene.
    (c) Cumene then anisole.

(3) Hexane (bp = 68 °C), 1-methyl-1-(1-methyl-ethoxy) ethane (diisopropyl ether) (bp = 68 °C) and propan-1-ol (bp = 83 °C), on polyethylene glycol succinate (PEG-S).

    (a) Hexane, diisopropyl ether, then propan-2-ol.
    (b) Diisopropyl ether, hexane, then propan-2-ol.
    (c) Hexane, propan-2-ol, then diisopropyl ether.

SAQ 3.7e

While the foregoing discussion should provide you with the knowledge upon which to make your decision as to the suitability of a column for a particular separation, in the laboratory you will frequently be expected to provide rapid analysis of unknown and sometimes complex mixtures. In this environment you need to have at your disposal a limited number of good quality and reliable columns.

Generalisations are often dangerous, but I would suggest that the two most important columns in your armoury should be OV-1 and Carbowax 20M. These columns should be say 5 ft long, of 4 mm internal diameter and contain 100–120 mesh support coated with 5–7.5% stationary phase. Although they are not at the extreme ends of the polarity range, you will probably find that these columns will give sufficient discrimination to enable you to complete a large proportion of your work efficiently. If you have two instruments, then you can have a 'polar' instrument and a 'non-polar' instrument. The less time you need to spend changing columns, the more analytical results you can produce.

OV-1 has a distinct advantage over non-polar stationary phases such as squalane and Apiezon in having much greater temperature stability. Indeed, the OV phases are, to all intents and purposes, the modern day equivalents of the early phases. OV-225 is a useful alternative to Carbowax 20M, again with a higher operating temperature limit.

If you want to produce some interesting results you might consider mixed-phase columns. I found that a 50:50 mixture of OV-1 and OV-225, the column packings having been mixed thoroughly before packing, gave useful, although not always predictable separations. You must, however, not exceed the maximum temperature of the most volatile phase.

3.8. SOLID STATIONARY PHASES

In general, these have been much less popular than liquid stationary phases because they frequently lead to long retention times, badly tailed peaks and poor reproducibility. Nonetheless, I find them quite interesting because they show selectivity amongst the components of a mixture which is far in excess of anything of which liquid stationary phases are capable. By and large, you are most likely to use them for separating gases — something which cannot easily be done with liquid stationary phases. This is a somewhat more specialised area of gas chromatography than the analysis of organic liquids, but it is still an important application.

Solid stationary phases can be loosely divided into two groups: those which operate by adsorption and those which operate on the *molecular sieving* principle. Naturally, it is not as clear cut as that, and some molecular sieves also act by adsorption, but it is convenient to treat the two separately.

Adsorption onto a solid surface is not a uniform process. Some areas of the surface adsorb molecules more readily and more strongly than others and it is in these areas that the adsorbed molecules cluster most thickly. Such areas are known as *active sites* and they tend to be found at physical irregularities (cracks and crevices, etc.) and at crystal lattice defects on the surface. It is this non-uniformity which leads to non-linear adsorption isotherms and tailing, as explained in *ACOL: Chromatographic Separations*. If you feel that you need to re-familiarise yourself with this you should read the relevant section in that unit before proceeding.

The forces responsible for adsorption are remarkably similar to those responsibilities for solubility — hydrogen bonding, dipole–dipole attraction, dipole-induced dipole attraction, dispersion forces, etc. In the context of gas chromatography you may find that acid/base interactions (the donation of a lone pair from an atom of one molecule to an electron deficient atom in another) assumes a greater importance, but otherwise they are very much the same. The extra selectivity of adsorption, compared with solution, probably arises from the rigid geometry of the solid surface. The atoms of the adsorbent which are responsible for attracting the adsorbed molecule are held rigidly in place by their crystal lattice. Their ability to attract the relevant atoms of the adsorbed molecule will depend upon the geometry of that molecule much more than in the case of attractions to the flexible, movable liquid molecules.

Molecular sieves, at least to some extent, function because of the many small molecular sized pores which penetrate their structures. Small molecules are able to penetrate into these pores and become adsorbed on their inner surfaces. Larger molecules are less able to penetrate, becoming adsorbed on the outer surfaces and so are eluted more quickly. The order of elution from columns of such stationary phases is roughly the order of decreasing molecular size or relative molecular mass. Of course it is never that simple; molecular sieves are also adsorbent and the differences in ease of adsorption of the molecules will be superimposed on this pattern. Still, you can manage quite a number of otherwise impossible separations with molecular sieves.

Common solid stationary phases are:

Alumina

Alumina (Al_2O_3) is a powerful adsorbent. It can hydrogen bond through hydroxyl groups formed on its surface by hydration, attract by dipole–dipole and dipole-induced dipole attractions through the same groups and through oxygen atoms in the surface, and electron deficient aluminium atoms can accept electron donation (acid/base interaction). It does an excellent job of separating the lower alkanes and alkenes and of analysing mixtures of freons. However, components are very susceptible to tailing and although this can be partly overcome by depositing inorganic salts onto its surface, you can never get truly symmetrical peaks.

Carbon Black

Carbon black has long been used as an adsorbent for GSC. In principle it should have no polar groups in its surface so that adsorption would be due solely to dispersion forces. The possession by the components of the mixture of functional groups, π-bonds or lone pairs of electrons, is therefore irrelevant to their adsorption. This should be controlled by the size, shape and polarisability of their molecules, and so, although selectivity between molecules could still be very high, it should be much less species dependent.

In practice, simple carbon black does possess a variable number of polar groups on its surface and these lead to badly shaped peaks and unreliable performance. Not surprisingly, carbon black in this form was used only sporadically. If, however, carbon black is heated in an inert atmosphere to about 3000 °C many impurities are removed and the surface undergoes a transformation into graphite. Such graphitised carbon black (GCB) is a much improved adsorbent. Even so, there are still physical irregularities and a few small areas of polar groups in the surface which have the effect of deforming peak shapes. This can normally be largely overcome by coating the adsorbent with a small amount (up to 1 or 2%) of a liquid stationary phase (often PEG). Whether you should think of this as just blocking the active sites or whether you should now think of the mechanism as a combination of partition and adsorbtion is arguable. What is significant is that vastly improved peak shapes and somewhat reduced retention times result and the stationary phase can be used to separate a wide range of organic compounds, both gases and liquids. These include saturated hydrocarbons, amines, phenols and aromatic acids.

Zeolites

These are the original alumino-silicate molecular sieves. They are also powerful adsorbents, adsorbing water, carbon dioxide, etc., irreversibly below temperatures of around 200 °C. (This means that columns must be protected from the atmosphere when not in use.) Of course, they cannot be used for the gas chromatography of these gases. They are, however, excellent for separating the noble (inert) gases, oxygen, nitrogen, carbon monoxide and other gases and for separating straight and branched chain hydrocarbons, all on the basis of size.

They are marketed by Union Carbide as the Linde range of molecular sieves.

Silica Gel

Silica gel is slightly odd in that, although it is porous, and its pore size certainly influences its performance as a stationary phase, it operates fundamentally as an adsorbent, not as a molecular sieve. Hydroxyl groups on the surface seem to be the main sites of adsorption; dispersion forces, dipole and induced dipole interactions and hydrogen bonding are all important, but there is some evidence that, in hydrogen bonding, it acts much more effectively as a proton acceptor than as a proton donor.

$$-Si-O\underset{`\text{H}-\text{X}}{\overset{\nearrow\text{H}}{\diagdown}}$$

It finds many uses in gas analysis, including the separation of carbon dioxide from other gases.

Porous Polymers

Styrene can be polymerised under conditions which lead to beads of porous, crosslinked polystyrene. Again, pore size affects chromatographic performance, but the mode of action is adsorption rather than molecular sieving. Adsorption takes place on the surface of the polymeric aromatic hydrocarbon within the pores, rather as it does on the surface of graphitised carbon black. The pores seem chiefly to ensure a very large surface area for adsorption to take place on. If anything, the polystyrene surface has even fewer extraneous polar groups than GCB, and peaks are very symmetrical. You can get excellent chromatograms of such components as water, ammonia, the lower alcohols and amines, all with very little sign of tailing.

They are marketed by Waters Associates under the name Porapak and by Johns-Manville as Chromosorbs.

Tenax-GC is another porous polymer which is available in bead form for packing gas chromatography columns. It is based upon 2,6-diphenyl-p-phenylene oxide and has a greater thermal stability than polystyrene. Slight column bleeding is noticed above 320 °C, but this does not become excessive until a temperature of 375 °C is reached. Like polystyrene, it gives very symmetrical peaks and with its high temperature limit it is excellent for separating higher boiling polar compounds such as aromatic amines, phenols, glycols, etc.

Perhaps the best way to get some idea of the separations that are possible by gsc is to look at the following chromatograms. They *say* much more than pages of writing. They were all obtained, using a routine instrument fitted with conventional, packed columns (1.5 m × 4 mm ID) and a carrier gas flow rate of 45 cm³ min⁻¹. Figures 3.8a, b and c were obtained using a katharometer or thermal conductivity detector (TCD) and helium carrier gas; Fig. 3.8d was obtained using an FID and nitrogen carrier gas.

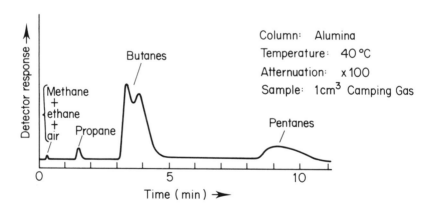

Fig. 3.8a. *Chromatogram of camping gas*

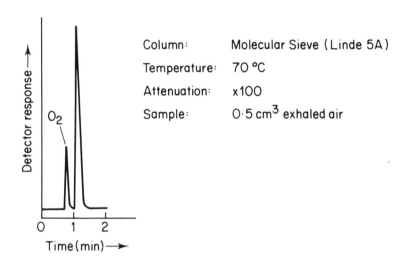

Fig. 3.8b. *Chromatogram of exhaled air — I*

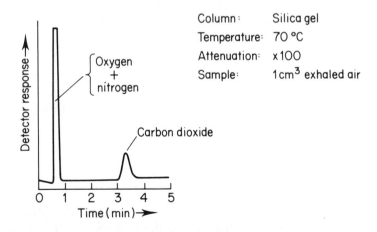

Fig. 3.8c. *Chromatogram of exhaled air — II*

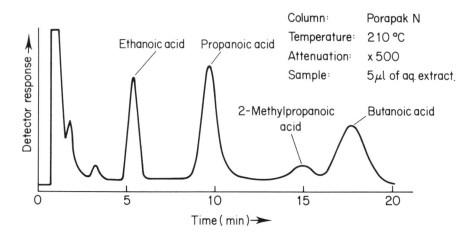

Fig. 3.8d. *Chromatogram of the alkanoic acids extracted from raw sewage sludge*

Π Would you expect benzene (bp = 80 °C) to elute before, at the same time or after cyclohexane (bp = 80 °C) on an alumina column?

If you answered 'after', well done. If you thought it would elute before or at the same time, then you had probably forgotten that the very polar groups in the surface of alumina would induce a dipole in benzene by polarising its π-electron cloud. This would cause it to be more strongly adsorbed than cyclohexane, which has only much less polarisable σ-bonds.

Π Which of the solid stationary phases that we have discussed would you choose to separate the following mixtures?

(1) Methane, ethane, propane, butane.

(2) Helium, neon, argon.

(1) If you chose a graphitised carbon black, preferably modified with a little PEG 400, you have made the best choice. It will give more symmetrical peaks than alumina, which is the other phase capable of separating them.

(2) You should have chosen a Linde molecular sieve. The noble gases offer no interactions with other molecules which you can use to differentiate between them — you can exploit only their size differences.

SAQ 3.8a Explain, in the space below, why gases are more likely to be analysed by GSC than GLC.

SAQ 3.8b

> If you were asked to determine the concentration of carbon monoxide and carbon dioxide in a boiler flue, which of the following stationary phases would you use?
>
> (1) To determine carbon monoxide.
>
> (2) To determine carbon dioxide.
>
> Stationary phases:
>
> Alumina, silica gel, graphitised carbon black, Linde molecular sieve, porous polystyrene (Porapak).

In summary liquid stationary phases are in much more common use than solid stationary phases, probably because of their great reliability. Liquids that are to be used as stationary phases must be non-volatile and chemically and thermally stable. They are best classified in terms of their polarity, that is as *non-polar, intermediate, moderately polar* and *very polar*. Their ability to dissolve and retain components correlates quite well with the similarity between the polarity of the compound being chromatographed and the stationary phase. For finer tuning and a more sophisticated choice of stationary phase it is necessary to look more deeply at the reasons underlying solubility and to consider the main attractions between solvent and solute molecules. These are hydrogen bonding, dipole–dipole attractions and dipole-induced dipole attractions, with dispersion forces making a lesser contribution.

Solid stationary phases, whilst being less commonly used, are attractive because they offer much greater selectivity and in fact, are often essential for the analysis of gases. They are, however, easily contaminated and retention times and separations may not be very reproducible. Selectivity will be due to differences in hydrogen bonding ability, dipole and induced dipole interactions and to dispersion forces or to differences in molecular size.

3.9.   CARRIER GASES

The choice of gas to be used as mobile phase in gas chromatography is influenced by the following requirements and considerations:

   inertness
   dryness
   freedom from oxygen
   safety
   cost
   availability.

Unlike capillary column gas chromatography, the molecular weight of the carrier gas is unlikely to have any noticeable effect on your chromatography when using packed columns.

Nitrogen is perhaps most widely used, in some laboratories being taken from a cryogenic nitrogen supply, thus eliminating the need to handle heavy gas cylinders. This source has the added advantage that it is unlikely to run out in the night even if you have a significant leak. Cryogenic nitrogen is very pure and contains such low levels of oxygen that there is no risk to the column.

Argon was widely used in the days of the argon ionisation detector and remained in use long into flame ionisation detector days due to its high purity and availability.

There are those of us who have used hydrogen as a carrier gas but really the explosion risk is not worth the worry and there is no significant performance advantage to be gained in packed column GC.

In recent years carbon dioxide has come into use in *supercritical fluid chromatography* (SFC). This technique uses supercritical CO_2 as the carrier and may be visualised as dense gas chromatography. Carbon dioxide is a more polar gas than the gases normally used in gas chromatography and therefore SFC falls into a category between GC and HPLC. SFC must be the subject of another book.

Gas control systems were discussed in Part 2.

3.10. THE INJECTION SYSTEM

A gas chromatograph is a sophisticated and sensitive scientific instrument and you should therefore expect that the sample introduction technique is important and that the hardware for sample injection should be robust and reliable. It is nevertheless amazing how often an apparent 'total failure of the chromatographic column' or 'the detector has died' or 'this recorder is duff!' syndromes, turn out to be simple injection problems, resolvable in minutes.

It is not that there is anything inherently wrong with the designs and construction of injection systems and syringes, it is just that we do not look after them and give them the attention they merit. We are dealing with very small quantities of solutions and materials, so care and developed skills are needed for success.

Injection systems for packed column gas chromatography are almost exclusively based on syringe injection through a septum. This is true for injection of solutions and also for gas sample analysis. Figure 3.10a shows the construction of a typical injection head to which the packed column is connected. There are other variants which include preheating of the carrier gas, but fundamentally they are all very similar.

There are a number of schools of thought regarding the heating of the injection head and the region of the column into which the

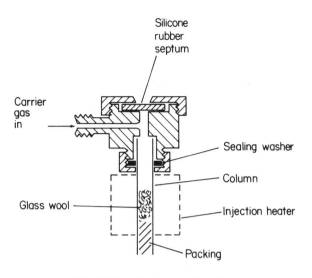

Fig. 3.10a. *Injection head*

syringe discharges its sample. Some workers favour injecting into a heated void above the column packing in which volatilisation takes place immediately, with solvent and solute vapours passing on to the column packing. For thermally unstable materials this can result in decomposition of solutes, resulting in confusing chromatograms and poor quantitative analysis. Others inject into the glass wool plug or packing material held at the maximum operating temperature of the stationary phase. If instrument construction permits, I prefer to inject into column packing inside the column oven. Since we are endeavouring to establish an equilibrium between the gas phase and the stationary phase, this is the best way to achieve such conditions with the added benefit of not subjecting thermally labile solutes to unnecessary thermal shock. This practice is usually possible with manual injection using long needle syringes. However, if you are fortunate enough to have an automatic injector this will almost certainly use a shorter needle syringe and inject into a heated zone outside the column oven. In this case you must compromise and run the injector zone at a temperature hotter than the initial oven temperature, since if you are temperature programming your elution, you must ensure that all solutes move into the column oven. Try some initial runs on your mixture at different injector zone temperatures to ensure that you are not discriminating against the less volatile solutes.

Early chromatographs frequently had 'cool' spots between injector and detector heated zones and the column oven and, to compensate, operators used to run the zones at a temperature hotter than the column oven to prevent solute hold-up. These problems no longer exist, so keep zone temperatures at sensible levels.

The injection septum forms a resealable access for the syringe needle. Septa are generally made from silicone rubber, however they will degrade at prolonged high temperatures, resulting in baseline drift and even additional peaks. PTFE-faced septa are available which largely overcome this problem but do remember to fit them with the PTFE face towards the column packing. Obviously septa have a limited life after which they will start to leak, resulting in loss of variable amounts of sample as vapour. Check septa frequently for leaks with a moist finger, but do take care as the septum cap may be rather hot.

Figure 3.10b shows the construction of a typical microlitre syringe. This is a piece of precision engineering and should be treated as such. Void volumes have been eliminated by having the plunger as a closely fitting wire of the same diameter as the internal diameter of the needle. Excellent care and maintenance instructions are provided by the syringe manufacturers, so do read them carefully. Always flush your syringe with solvent after use to avoid crystallisation of solute and seizing of the wire inside the needle.

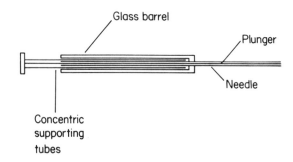

Fig. 3.10b. *Microlitre syringe (capacity up to 1.0 μl)*

∏ Which of the following is correct?

    (a) $10\,\mu l = 0.1\,ml$

    (b) $10\,\mu l = 10^{-3}\,ml$

    (c) $10\,\mu l = 0.01\,ml$

    (d) $10\,\mu l = 10 \times 10^{-6}\,ml$

The correct answer is (c), $10\,\mu l = 0.01\,ml$.

By developing a technique whereby you operate the syringe with one hand only, the risk of overdrawing the central wire out of the needle and subsequent crumpling can be avoided. Alternatively an adapter can be fitted which restricts the travel of the plunger. Syringes of greater than $1\,\mu l$ capacity are based on a closely fitting metal plunger inside the glass barrel as are gas syringes, albeit with a larger volume and a gas tight seal incorporated into the plunger (Fig. 3.10c).

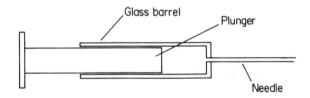

Fig. 3.10c. *Gas syringe (capacity up to 25 ml)*

It is good practice to check the tip of the needle for any raggedness which may tear the septum. Polish off any such defects with fine emery paper.

When using the syringe, first flush it several times with sample solution, discharging the flushings onto a tissue to avoid the risk of sample cross contamination. Then fill the syringe, remove it from the sample vial and carefully wipe the needle with a tissue. Move the plunger to the desired volume, say $0.2\,\mu l$, and quickly and lightly wipe the needle again to remove any excess liquid. Do not allow the liquid to be lost by capillary attraction of the tissue. Pass the needle through the septum into the column and depress the plunger smartly. Pause for a second or so before withdrawing the needle fairly slowly from the septum. Slow withdrawal allows the septum to reseal before the syringe needle is completely removed thus preventing sample vapour being lost and the gas flow conditions inside the column being disturbed.

The accuracy and precision of your gas chromatographic analysis will be very much a reflection of your skill and attention to detail in injection. Remember of course that as the injection volume decreases, the precision of the particular syringe also decreases. If you attempt to inject 0.05 μl using a 1 μl syringe, then repeatability of the resulting peak areas will be poorer than if you injected 0.5 μl. Even with a good technique, it will be difficult to overcome variable evaporation of the solvent and possibly solutes from the very small volume of liquid at the tip of the needle. Better results at the 0.05 μl level will probably be achieved using a 0.5 μl syringe but the more appropriate solution (pardon the pun!) is to dilute the sample solution by a factor of ×5 or ×10 and inject 0.5 or 1 μl. At these volumes the syringes are more accurate and repeatability should be enhanced.

3.11. SYSTEM OPTIMISATION

Towards the end of Section 3.7 it was suggested that OV-1 and Carbowax 20M columns might be your work-horses for most of your separations.

Further optimisation of separations involves the utilisation of the other chromatographic parameters at your disposal, namely column length, temperature and temperature programming, injection volume, sample solution concentration and carrier gas flow rate. Attenuation or 'sensitivity' do not have anything to do with separation, they determine whether or not you are going to see the detector response to solutes passing through.

Increasing the column length will increase the number of theoretical plates available for your separation, although at the expense of analysis time. Do not be afraid to try a shorter column if there appears to be more than adequate resolution. Analytical productivity will be improved by reducing the analysis time and increasing sample throughput.

Optimisation of temperature is very much a matter of experience. For a single compound analysis (if it is really justified!) or for the separation of a couple of closely eluting compounds, select a temperature about 20 °C below the boiling point. If you are to separate a more complex mixture of compounds then a fairly wide-ranging temperature programmed run will ensure that all the components stand a reasonable chance of elution. Once you are sure you are accounting for everything, then you may establish a more optimum programme range or rate or even isothermal conditions. Long periods of flat baseline between peaks is simply wasting analysis time.

Separations may also be improved by injecting a smaller sample volume or even a dilute solution. Injections of neat material may result in a degree of overloading and band broadening which may make resolution of closely eluting peaks more difficult.

From your study of *ACOL: Chromatographic Separations* you will know that at an optimum carrier gas flow rate, maximum column efficiency is achieved. Generally this is in the range 30–50 ml/min. It is a good exercise to carry out the determination and to generate an *HETP* curve. Analysis time can be reduced by increasing the flow rate and, in cases where compounds are exhibiting a degree of thermal instability, this may be a better approach to reducing the analysis time than increasing the oven temperature. Increasing flow rate can also be used to reduce column overload effects.

Finally, having achieved a separation and analysis time which suits your purpose, do make sure to record all the conditions accurately. Rubber stamps are available as memory joggers to ensure that vital information is not missed, maybe an out-of-date practice but effective nonetheless. On computer driven systems, operational parameters are entered through the keyboard and can be recalled and viewed on the screen. With some software it is possible to make access for additional information, e.g. column type and dimensions etc., and to store all the information under a suitable method name.

If you can go back at a later date, set up your column or, better still, an identical column with the established conditions and obtain satisfactory analysis straight away then you can be satisfied that you have done a good job in establishing a robust method.

3.12. ASSESSMENT OF CHROMATOGRAPHY

It is impossible to provide examples of all of the types of chromatographic separation and performance which you are likely to come across in your work. It is possible, however, by means of simulated chromatograms, to illustrate some of the problems which may arise which would adversely affect the quality of your data and to suggest how the situation may be improved.

Resolution, i.e. the separation between peaks in the chromatogram is usually the first concern and if the analysis can be achieved in a reasonable time-scale, so much the better.

Figure 3.12a is quite characteristic of a first-attempt chromatogram. Peaks are sharp but unresolved and analysis time is very short.

∏ Suggest how the chromatography might be improved.

The first action would be to reduce the temperature. It would also be prudent to check that the column flow rate is somewhere near the

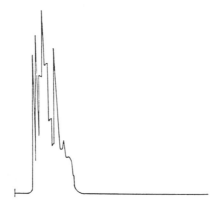

Fig. 3.12a. *Peaks unresolved; appearing in a very short time*

optimum. At first sight this looks like a fairly complex mixture so a longer column, a heavier loading of stationary phase or even temperature programming may prove necessary to produce the 'perfect' separation.

Figure 3.12b shows a lack of resolution but a reasonable analysis time. It is unlikely that this separation could be significantly improved on this column but perhaps a different stationary phase or use of capillary column may provide a more suitable alternative.

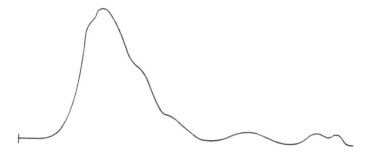

Fig. 3.12b. *Resolution poor; analysis time reasonable*

Figure 3.12c illustrates a potential separation still in need of further work. Analysis time is reasonable but the peaks are rather broad indicating that the column efficiency is not too great, although peak shape is reasonable.

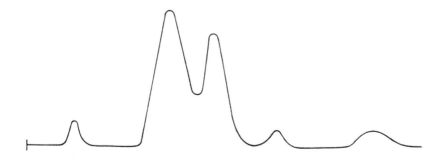

Fig. 3.12c. *Resolution inadequate, but promising*

∏ Which parameters should be considered to improve this
   separation?

You must identify what is the cause of the broadened peaks. The
optimum solution probably involves the combination of a longer
column, lower stationary phase loading and a slightly higher
temperature. If, however, the broadened peaks are due to heavy
sample loading then a smaller injection volume or a more dilute
solution could improve the chromatography without recourse to the
other possibilities.

Figures 3.12d(i) and (ii) show chromatograms with asymmetric peaks
typical of overloading. These are rather extreme examples of peak
asymmetry. It is unusual to get completely symmetric peaks in
chromatography, as peaks generally show a slight degree of tailing due
to adsorption on residual active sites. The chromatogram shown in (i)
is typical of overloading on porous polymer bead columns and can be
improved by reducing the sample loading.

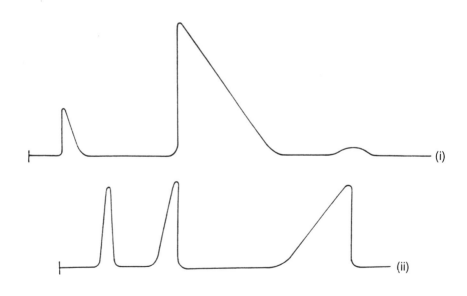

Fig. 3.12d. *Asymmetric peaks; two types*

Figure 3.12d(ii) is typical of overloading of partition columns containing conventional stationary phases. There are four possible solutions.

(a) Reduce the injection volume or the sample solution concentration.

(b) Increase the stationary phase loading of the column to provide a greater sample capacity.

(c) Increase the internal diameter of the column.

(d) Increase the carrier gas flow rate.

Peak tailing shown in Fig. 3.12e, as discussed earlier, arises due to adsorption of the solute by active sites in the column or in the connecting pipework. In severe cases it may be necessary to prepare a new column using a more inert solid support. This may be a different material or simply the same support which has been further treated by silanisation to reduce the active sites.

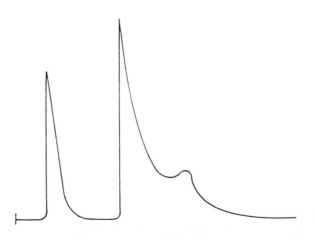

Fig. 3.12e. *Peak tailing*

∏ Why is it essential that we solve the problem of asymmetric peaks?

Peak tailing and peak symmetry can adversely effect quantitative analysis. In the case of overload peaks the peak is not running to its potential maximum amplitude and this will result in a lower integral count. If minor components are also present in the mixture then these will be overestimated if analysis is carried out on a peak area normalised basis.

Thermal decomposition shows up in gas chromatography as a rising baseline leading up to the peak but the most characteristic feature is that the baseline returns to its original level immediately after the peak has been eluted, Fig. 3.12f. Deactivation of the column or reduction of the column temperature and the injection temperature (if separately controlled) may ease the problem. It is possible that thermal decomposition may be avoided by using a capillary column and cold on-column injection.

Fig. 3.12f. *Peak leading; thermal degradation*

Figure 3.12g is typical of many chromatograms you will see. Resolution is more than adequate and the analysis time could be significantly reduced. Reducing the analysis time will sharpen the later peaks and improve quantitative accuracy, particularly if some of the later peaks are small.

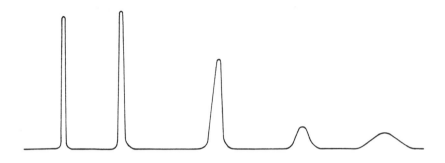

Fig. 3.12g. *Long analysis time*

∏ How would you reduce the analysis time?

You will by now realise that there are a number of ways of reducing the analysis time, namely, higher oven temperature, higher carrier gas flow rate, shorter column or a column containing less stationary phase.

Figure 3.12h shows a chromatogram in which the column temperature was too high to allow adequate separation of the early components in the mixture and too low for the later peaks which are drawn out and broadened. Here temperature programming provides the ideal solution. The initial temperature will be chosen to optimise the separation of the early peaks and then the temperature programmed at such a rate so as to produce sharp and resolved peaks. Ideally, peak-width should remain virtually constant through a temperature programmed chromatogram. Do not expect to get it right first time, it will probably need a number of attempts to get the best results.

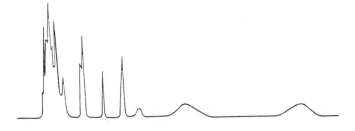

Fig. 3.12h. *Long analysis time; early peaks unresolved*

When looking at a new sample or a new sample mixture it is good practice to carry out a programmed run to ensure that you are seeing all the volatile components in the mixture. It can be rather embarrassing to have given out your results only to discover that additional peaks come out later.

Don't be afraid to *reduce* the attenuation setting, i.e. increase the sensitivity, to ensure that you are seeing all the components on the chromatogram.

The temperature programme may result in baseline drift (Fig. 3.12i) which may or may not make quantitative analysis more difficult. Some instruments have baseline compensation built in and this can assist considerably. The alternative may be to choose a more thermally stable column and stationary phase or programme to a lower final temperature.

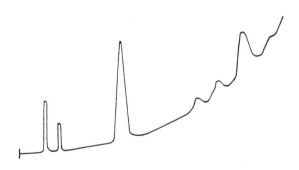

Fig. 3.12i. *Baseline drift*

In some cases, conditioning the column overnight near its maximum operating temperature may also reduce the drift.

I believe that chromatography is an art as well as a science. If the chromatography looks good then it probably is good.

Learning Objectives of Part 3

After studying the material in Part 3, you should be able to:

- describe a packed gas chromatography column;

- describe the preparation of packed columns;

- discuss the mechanism by which common liquid stationary phases retain the components of a mixture;

- choose the column dimensions, temperature, flow-rate, sample size and attenuation appropriate for a given analysis.

4. Capillary Column Systems

4.1. HISTORY OF CAPILLARY COLUMNS

It would be easy to come to the conclusion that capillary columns were a very recent innovation in gas chromatography since it has only been in the past 10 years that capillary column gas chromatography, now more commonly referred to as high resolution gas chromatography (HRGC), has made a significant impact on the chromatography market.

This belies the fact that the invention and the study of the principles of capillary column chromatography dates back to 1958 and the work of Marcel Golay of Perkin-Elmer. Capillary columns were frequently referred to as Golay columns.

It is worth considering Golay's rationale for developing capillary columns. While studying the effect of phase loading on packed column performance, Golay postulated that the pathways taken by the gas phase as it passed through the column between particles were equivalent to several open capillaries.

Following this realisation we can read in one of Golay's reports 'it is suggested that experiments be undertaken with a single capillary, 0.05 or 0.1 cm in diameter, wetted with a silicone oil, for the sake of determining to what extent the theoretical possibilities expressed earlier may be realised'.

This was the start. Early experiments were rather disappointing due to the inadequacies of the designs of injection systems and detectors. Golay persisted and in 1958 at the Amsterdam Chromatography Symposium presented his theory of capillary column chromatography and examples of successful separations. Most remarkable of these was the separation of *meta-* and *para*-xylene, a first in those days.

SAQ 4.1a

> Golay's expectations were not realised in his initial experiments in capillary chromatography. What do you think the inadequacies of the injection systems and detectors were that are referred to in the text?

In many respects, the development of capillary columns was inhibited by the patenting of the technology by Perkin-Elmer. However, developments in column technology and coating techniques, principally using glass columns, continued with particularly notable success by the Grob family in Switzerland.

Glass columns were never easy to use, required delicate handling and were not really ideal for routine laboratory use. Problems with discriminating injection systems further hampered acceptance.

The development of fused silica columns with chemically bonded phases in the 1980s marked a breakthrough in column technology and a resurgence of interest in gas chromatography.

4.2.   COLUMN STRUCTURE AND DIMENSIONS

Capillary columns can be divided into two types:

WCOT   wall coated open tubular columns
SCOT   support coated open tubular columns
        and related PLOT columns,
        (porous layer open tubular columns)

Wall coated columns are by far the most common, comprising a glass, silica, or steel tube with an internal diameter in the range 0.1–0.5 mm. Figure 4.2a shows the cross-section of a present-day bonded phase fused silica column.

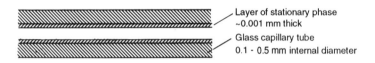

Layer of stationary phase
~0.001 mm thick

Glass capillary tube
0.1 - 0.5 mm internal diameter

Fig. 4.2a. *Glass capillary column*

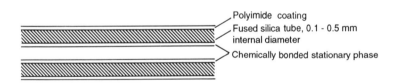

Polyimide coating
Fused silica tube, 0.1 - 0.5 mm
internal diameter
Chemically bonded stationary phase

Fig. 4.2b. *Fused silica bonded phase column*

Cross-sections of SCOT and PLOT columns are shown in Figs 4.2c and 4.2d, respectively.

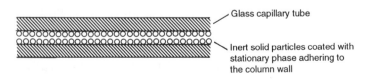

Glass capillary tube

Inert solid particles coated with
stationary phase adhering to
the column wall

Fig. 4.2c. *Support coated open tubular column (SCOT)*

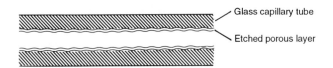

Fig. 4.2d. *Porous layer open tubular column (PLOT)*

The early wall coated capillary columns were prepared by passing a solution of the chosen stationary phase, dissolved in an organic solvent, through the column and then blowing it dry in a stream of inert gas. The quality of the internal surface of the glass capillary was important and many processes were developed to leach out metal ions from the glass surface prior to coating since it was discovered that the presence of these impurities gave rise to adsorption problems and unsatisfactory peak shapes.

Fused silica columns use a core of very high purity silica, virtually free of metal ions. However, this silica is susceptible to atmospheric oxidation which causes breakage. This weakness is overcome by coating the column with a polyimide which protects the column, retains the flexibility of the fused silica and makes it easy to install and to handle.

Π Why do you think the problems of fragility in glass capillary columns could not be overcome by coating with a polyimide in the same manner as fused silica columns?

Perhaps they could, but the function of the polyimide coating on fused silica columns is to protect the silica from atmospheric degradation, not to make the column more flexible. Generally, glass columns were made from thicker wall tubing which made them much less flexible. You will remember that one of the problems with glass columns was the metal ion content of the glass which could give rise to poor chromatographic performance and required considerable leaching to render the material more suitable. Fused silica is very pure, essentially free from metal ion contamination, and is therefore more suitable for column manufacture.

The stationary phase is chemically bonded onto the silica and a wide range of polarity is now available. The chemical processes involved in the bonding and cross-linking of stationary phases on capillaries is beyond the scope of this text. However, interested readers are recommended to read the book entitled *Making and Manipulating Capillary Columns for Gas Chromatography* by Kurt Grob for a fund of information. When compared with film coated columns these bonded phase columns have a considerably higher maximum operating temperature, low bleed and a much longer useful life. In addition, by washing the column with an appropriate range of solvents it is frequently possible to recover the performance of a degraded column.

Π Suggest how the stationary may be bonded to the capillary column wall?

Silica tubing has many hydroxyl groups, i.e. silanol groups, to which may be attached a suitable silane; part of the stationary phase molecule is shown below:

$$\text{Si} {-}\text{OH} + \text{Cl} - \underset{\underset{R}{|}}{\overset{\overset{R}{|}}{\text{Si}}} - R^1 \longrightarrow \text{Si} {-}\text{O} - \underset{\underset{R}{|}}{\overset{\overset{R}{|}}{\text{Si}}} - R^1$$

Film thickness is important because it affects both the phase ratio, the ratio of stationary phase to mobile phase which controls retention times and separations, as well as the maximum sample loading that the column can handle. For a given component, the thicker the film, the longer the retention time. Film thicknesses range from 0.1–$1.0\,\mu m$.

Internal diameters range from 0.1–$0.5\,mm$. Smaller bore columns with an internal diameter of $0.05\,mm$ are manufactured for use in supercritical fluid chromatography.

Columns with an internal diameter (i.d.) of $0.1\,mm$ are recommended for use with bench top mass spectrometers while for general use $0.25\,mm$ or $0.32\,mm$ i.d. columns are most commonly used. Wider bore columns, often referred to as 'Megabore' columns, are recommended as replacements for packed columns. I personally have not found it that simple, but they undoubtedly have a role as I shall demonstrate later.

Column lengths range up to 60 m but 15–30 m are more common. We all generally work with columns much longer than we need. Professor Walter Jennings, one of the authorities on HRGC, recommends that we purchase a 30 m column, cut off one-third of the column and use that for our analysis. He maintains that in most instances that will do for our analysis and we are then left with an additional 20 m column or two 10 m columns for later use. I have come across few chromatographers who follow this advice. It takes some courage to cut up a £300 column. If you decide to try this approach, do make sure you have plenty of room to work, unwind your column carefully and lay it out as you go along, otherwise you may end up with the most awful tangle on your hands.

∏ What could you use to provide you with a fast separation column which could be run at normal temperatures and flow rates to assess possible methods quickly?

You could have suggested a 5 m length of column taken from a 30 m column. This would certainly be fast and a loss of 5 m is not going to be noticed from the resolution performance of the remaining 25 m length.

It is unlikely that you will ever use the second category of column, namely the SCOT or PLOT columns, which predate the fused silica column, so discussion of these types will be brief.

SCOT or PLOT columns were developed in an attempt to achieve a high phase ratio in a wide-bore capillary without increasing the average film thickness. The aim was to allow high flow rates and large sample sizes without the loss of resolution which would result from a thick film. Unfortunately resolution never matched that obtained from conventional wall coated capillaries.

An internal porous layer was generated *in situ* by chemical etching of the internal glass surface prior to coating in the conventional manner, hence PLOT column. The alternative method, the SCOT column, was to fill the capillary with a stable suspension of 0.01 mm diameter support particles in a solution of the stationary phase in a volatile dense solvent. Passing a small zone furnace along the length of the capillary evaporated the solvent, depositing the support and stationary phase on the wall of the column.

In most respects the bonded phase 'Megabore' column mentioned earlier fulfils the role intended for the PLOT and SCOT columns.

SAQ 4.2a   Describe, with the aid of a labelled diagram, the construction of a capillary column which you might buy today from a laboratory supplier. How does it differ from a column manufactured in the 1970s and what are the advantages of the modern product?

SAQ 4.2a

4.3. RANGE OF BONDED PHASE CAPILLARY COLUMNS

Much of the discussion of stationary phase characteristics in Part 3 is relevant also to capillary columns.

There is a considerable range of capillary columns available from many manufacturers and since each manufacturer prefers their own classification and nomenclature, it might appear that the choice would be enormous. This is not really the case as many of the columns are equivalent, to all intents and purposes.

For example J & W Scientific Products DB-1 is a simple methylsilicone phase column. The following are equivalent — HP-1, HP-101, Ultra-1, SPB-1, CP-Sil 5, RSL-150, RSL-160, BP-1, CB-1, OV-1, PE-1, SP-2100 and SE-30. There are probably more.

I am not aware of any 'bad' columns and I believe that, in many respects, the performance you get from a column is more dependent on the way in which you look after it and the samples you analyse rather than any particular manufacturing process detail.

Table 4.3a.   *Bonded phase capillary columns*

Phase	Film thickness (mm)	Temperature range (°C)	Similar phases	Polarity
				Low ⫸
DB-1 Methylsilicone	0.10–1.00 3.00–5.00	−60 to 325/350 −60 to 280/300	OV-1, OV-101, SE-30, SP-2100, CP-Sil 5, DC-200, UCW-982, SF-96	
DB-1ht	0.10	−60 to 400		
DB-1 (Megabore)	0.15 1.50 3.00–5.00	−60 to 360 −60 to 300/320 −60 to 400		
DB-5/DB-5ms/DB-5.625 5% Phenyl	0.10–1.50 1.00	−60 to 325/350	SE-54, CP-SIL 8, SE-52, Dexsil 300, DC-200, Fluorolube, OV-73, DC-560, OV-3, OV-5	
DB-5ht	0.10	−60 to 400		
DB-5/DB-5ms (Megabore)	1.50	−60 to 300/325		
DB-1301 6% Cyanopropylphenyl	0.25–1.00	−20 to 280/300	none	
DB-1301 (Megabore)	1.00	−20 to 260/280		
DB-1701 14% Cyanopropylphenyl	0.05–1.00	−20 to 280/300	OV-1701, CP-Sil 19	
DB-1701 (Megabore)	1.00	−20 to 260/280		

High ▶

DB-35 35% Phenyl	0.15–0.50	40 to 280/300	Rtx 35, SPB-35
DB-35 (Megabore)	1.00	40 to 260/280	Rtx 35, SPB-35
DB-17 50% Phenyl	0.10–0.50	40 to 280/300	OV-17, SP-2250, OV-11, DC-710, OV-22, OV-25
DB-17ht	0.10	40 to 340/365	
DB-17 (Megabore)	1.00	40 to 260/280	
DB-210 50% Trifluoropropyl	0.15–0.50	45 to 240/260	OV-210, SP-2401, QF-1, UCON HB 280X, Triton X-100, OV-202, OV-215
DB-210 (Megabore)	1.00	45 to 220/240	
DB-225 50% Cyanopropylphenyl	0.15–0.25	40 to 220/240	OV-225, CS-5, XE-60, Silar 5 CP
DB-225 (Megabore)	1.00	40 to 200/220	XE-60, UCON HB 5100, AN-600
DB-WAX Polyethylene glycol	0.05–0.50	20 to 250/260	SP-1000, AT-1000, FFAP, OV-351, CARBOWAX 20M, Superox, PEG, DEGS, CPWAX 51, Supelcowax 10
DB-WAX (Megabore)	1.00	20 to 230/240	
DB-23 50% Cyanopropyl	0.15–0.25	40 to 250/260	OV-275, DEGS, SP-2310, SP 2330, CP-Sil 58
DB-23 (Megabore)	0.50	40 to 230/240	

Table 4.3a shows the range of stationary phases available using the J & W classification, together with the composition and the operating temperature ranges. Equivalences can be found in laboratory suppliers' or manufacturers' catalogues.

∏ What is the major disadvantage of polar capillary columns?

You could have deduced from your study of Table 4.3a that the more polar columns have lower recommended maximum operating temperatures, which is unfortunate.

SAQ 4.3a

Arrange the following bonded phase columns in order of increasing polarity: DB-17, DB-5, Ultra-2, DB-1701, HP-225, CP-Sil 5, SBP-1, HP-2, DB-WAX.

If starting out in capillary column gas chromatography, a column such as a DB-5 30 m × 0.25 mm internal diameter column, with a 0.25 μm film thickness, is a sound choice. You may well be surprised just how much of your analysis can be carried out on one column such as this. Columns are not cheap so choose wisely.

With experience and with very careful maturing of your columns you may find that you will be able to exceed the manufacturer's temperature recommendations by 10–20 degrees without any adverse effect. This can be worthwhile on occasions to extend further the range of your HRGC analysis.

In part, the temperature limitation of non-polar capillary columns is now the thermal stability of the polyimide protective film. Aluminium coated columns appeared on the market a few years ago and it was thought that this might be a way round the problem. Unfortunately, theory and practice turned out to be two different things and if you used the column, as I did, over a wide temperature programming range and frequent temperature cycles, the column soon started to break up into rather short sections. At a constant high temperature, however, I believe long-term performance is satisfactory.

It should be noted that the temperature limits of megabore columns, which generally have a thicker phase film, are lower than the smaller bore equivalents.

Finally in this section I shall mention a capillary replacement for the well known Porapak-Q packed column; this is GC-Q, designed for the analysis of low-molecular-weight alcohols, esters, hydrocarbons and sulphur gases. This is a gas–solid phase system which should simplify some difficult analyses.

4.4.   HANDLING CAPILLARY COLUMNS

In the early days of gas chromatography we used to make our own capillary columns. We used a glass drawing machine which produced coiled glass capillaries which looked like Slinkeys, the metal spiral toy which 'walks' down stairs.

These glass columns always seemed to be quite robust until they were coated with stationary phase. Then everything changed, the column would break as it was being fitted into the chromatograph or when couplings were being attached. They were not really any more fragile, it was just that the glass was becoming stressed as we handled it and made our gas connections. We also became stressed!

All this has now changed but do not be fooled into thinking that you need not exercise care in handling fused silica capillary columns. You must, and if you do, your column will give you excellent service.

Here are a few basic 'do nots':

- Do not leave your column lying on the bench out of its box for long periods. Use it or put it away.

- Do not physically stress the column when mounting it in the column oven. Allow the column to hang naturally from fittings and ensure the weight of the column is supported by the column frame.

- Do not allow the column to be exposed to the radiant heat from the oven heater.

- Do not switch on the column oven heater until the carrier gas has been flowing through the column for about 45 min.

- Do not run your column up to its maximum operating limit until it has been used for some time. Work up the temperature range progressively.

Now some 'dos':

- Do read and follow the instructions supplied with the column, they are for your benefit.

- Do use a proper cutter to cut your columns, the laboratory file will not do. Column cutting tools are available from laboratory suppliers and normally include useful tips and hints. Some workers use the silicone wafer from defective integrated circuits.

- Do slide on nuts and ferrules before making your cuts.

- Do check your cuts with a magnifying glass or better still with a microscope. If the cut looks ragged, shattered or broken then do it again. This is the most important stage of your installation so be prepared to take time. If you end up cutting off about a foot of column before you get a satisfactory break it does not matter and you will never notice the difference in your separation.

- Do remember to turn on the gas. It is probably better to attach the injector end first, turn on the gas, check for flow at the detector end and then connect to the detector.

- Do heed the chromatograph manufacturer's instructions regarding the lengths of capillary to be inserted into injectors and detectors. If the manual requires the column to be inserted 4 cm into the injector then measure 4 cm, mark it on the column with Tipp-Ex fluid, loosely connect the fitting and then slide the column up to the mark before tightening.

- Do check for leaks using an organic solvent, say heptane or ethyl acetate. Do not use soap solution, if you have a leak it may get into your column and give rise to odd peaks and baseline problems for some time. Apart from using solvents to check for leaks you can also check for them by observing the pressure drop in the system if the carrier gas supply is turned off as described in Section 2.4. This is a worthwhile test since you can use the method to check out each part of your chromatograph for leaks and could in the end save yourself considerable expenditure on helium. To check upstream of your column, fit a blank fitting to the column inlet connection. Pressurise the system to, say 2 bar, 28 psi, or 200 kPa, depending on your pressure gauge calibration. Now turn off the gas input to your system and observe the pressure reading. If it drops, then there is a leak somewhere in your system. Go through the gas flow system until you find the source of the leak and then fix it. Once the upstream end is gas-tight then connect your column. Let us assume that it is a 30 m × 0.25 mm i.d. column. Reduce the pressure setting to 1 bar, or the equivalent, pressurise the system and then turn off the inlet source once again and observe the pressure drop. This time there will be a drop in pressure as there will be a flow through the column but it should drop only slowly, say over a five minute period. Alternatively, you can place the detector end of the column into a silicone septum and then if everything is gas-tight there should be no pressure drop. Do remember to cut off the end of the column in case any septum material is left in the capillary when you remove the septum. The checks are worth the time in terms of protecting your column, saving on expensive carrier gas and ensuring system sensitivity.

Note. Ensure that the oven temperatures set are compatible with your column. Remember that injector and detector temperature settings should not exceed the recommended maximum column temperature.

SAQ 4.4a

> List four basic precautions to take when installing a capillary column in a gas chromatograph.

4.5. FITTINGS AND CONNECTORS

A wide range of fittings for joining columns, splitting flows, etc. are available from a number of manufacturers and a selection of these is illustrated in Fig. 4.5a.

Fittings can be divided into two types, those based on conventional compression fittings and glass press-fit connectors. Both types work very effectively but, as always, keep in mind the scale of your equipment and focus your attention to detail accordingly. Connections should be as near perfect as you can make them.

Compression fittings use vespel ferrules which can be purchased separately as required. Ferrules come drilled in different sizes to fit different column diameters. Place ferrules onto columns before making the cut to avoid a sliver of the ferrule material entering the column. Study the diagrams and information in the suppliers' catalogues, as these provide detail of the optimum assembly.

Glass press-fit connectors are a fairly recent innovation and in many respects are the ideal means of column connection. They are very light and therefore reduce any physical stress on the column. Different internal diameter connectors are available to suit their many purposes. The gas seal is achieved by adhesion of the polyimide layer of the column to the glass wall of the connector. This depends on the quality of the cut end of the column and it is essential to check for leaks after making a connection. If it leaks, try cutting again. Initially when making the connection the column can be removed without leaving any polyimide adhering to the glass. However, once the column has been up to temperature, the bonding is much stronger and almost certainly polyimide will be left on the glass if the column is removed. If in doubt have a look under the microscope. The presence of residual polyimide will almost certainly make it impossible to establish a gas-tight connection and even if you did you would have an adsorptive site within the column system which might well degrade your chromatography. In a sense, therefore, press-fit connectors should not be considered re-usable.

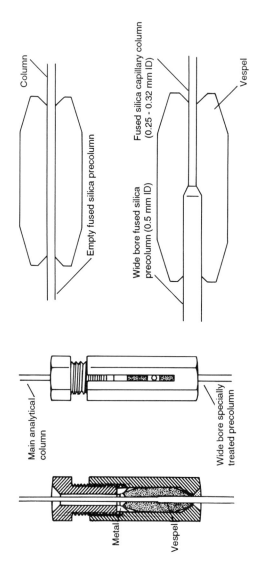

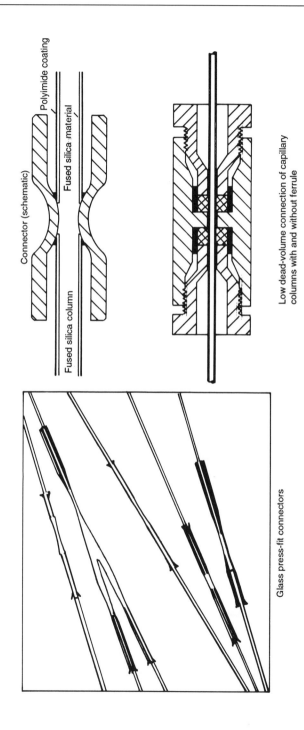

Fig. 4.5a. *Selection of fittings used for joining columns, splitting flows and other applications*

∏ List three key points to check when using connectors.

You could have chosen from the following:

- correct size fittings and ferrules

- column ends free from fractures or shattering

- no leaks

- no voids

- no stress on the column

- no ferrule material inside the column.

You should have had no difficulty thinking of three!

We shall discuss the use of multiway connectors later when we consider interfacing to mass spectrometers and other detectors.

A few words at this stage regarding *retention gaps*. A retention gap is a length of deactivated fused silica tubing containing no stationary phase used at the injection end of the column, particularly when using automatic on-column injection systems. Autoinjector syringe needles are made of steel and have to be robust enough to pass through seals and injection devices without bending. Their external diameter therefore exceeds the internal diameter of 0.25 mm and 0.32 mm columns. Retention gaps, 0.5 mm i.d. are therefore used connected to the analytical column by a reducing coupling to allow the syringe needle access to the column system, see Fig. 4.5b. More about on-column injection in Part 5.

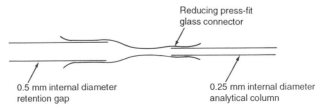

Fig. 4.5b. *Arrangement for connecting a retention gap to an analytical column*

The advantages of using retention gaps were not fully realised when they were first introduced as a solution to the automatic injection problem.

Retention gaps have made possible a technique for large-volume injection into capillaries which is particularly useful as a means of sensitivity enhancement for low-level component analysis. It also makes possible the direct combination of HPLC and HRGC, which we will consider in Part 8.

4.6.   CARRIER GASES IN HIGH RESOLUTION GAS CHROMATOGRAPHY

In packed column gas chromatography the nature of the carrier gas was not considered to be particularly important provided that it was clean, dry and free from oxygen.

SAQ 4.6a	Which of the following gases would be suitable as carrier gases in HRGC? Circle the appropriate response: yes (y), no (n).
	Argon y/n
	Hydrogen y/n
	Carbon dioxide y/n
	Air y/n
	Nitrogen y/n
	Helium y/n
	Xenon y/n
	Oxygen y/n

In capillary column gas chromatography there are performance advantages to be gained from using the lowest-molecular-weight carrier gas available.

From your study of the van Deemter equation in *ACOL: Chromatographic Separations* you will realise that the efficiency of a chromatographic column is related to the linear gas velocity, with the minimum of the curve corresponding to the maximum column efficiency, Fig. 4.6a. The minimum is termed the optimum gas velocity.

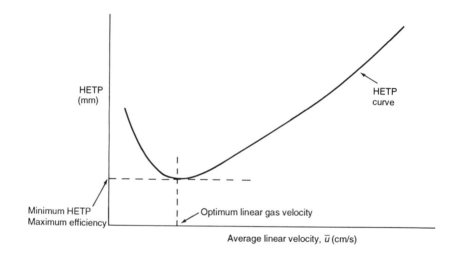

Fig. 4.6a. *Relationship between HETP and average linear gas velocity*

Π Quote the van Deemter equation. Which term becomes zero in capillary column gas chromatography?

If you had to go back to *ACOL: Chromatographic Separations* then that is fine. Knowing where to find information is an important part of learning.

You should have presented the equation as follows:

$$H = A + (B/\bar{\mu}) + C\bar{\mu}$$

where

$$A = \text{eddy diffusion term}$$

$$B = \text{longitudinal diffusion term}$$

$$C = \text{mass transfer term}$$

The eddy diffusion term, A, becomes zero in a capillary column since there are no particles around which to eddy, as the column is simply a straight tube.

If a series of van Deemter graphs are generated using the same column, the same solute but with nitrogen, helium and hydrogen as carrier gases and we overlay the plots, then we find that as the molecular weight of the gas decreases, the optimum gas velocity increases, Fig. 4.6b. Not only are the optimum gas velocities higher for the lighter gases but also the shape of the curve is considerably flattened.

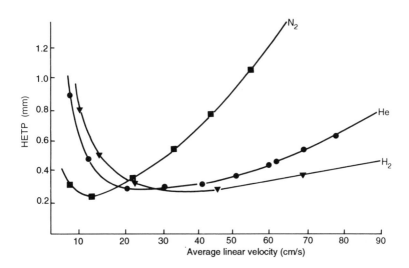

Fig. 4.6b *Comparison of HETP plots using different carrier gases (non-polar column, 25 m ×0.25 mm i.d.)*

The practical consequence is that, if we use hydrogen as carrier gas, maximum efficiency occurs at a higher gas velocity and over a wider gas velocity range.

The use of hydrogen as a carrier gas is not recommended although there are very convincing arguments, supported with numerical data, proving that the risk of explosion in a column oven is minimal in the event of a leak. If a hydrogen leak detector is fitted to the instrument it may be sensitive to other laboratory atmospheres and trip out the chromatograph at an inappropriate moment.

You might conclude that helium is the best compromise and providing you have cured all the leaks, you will find that a cylinder will last for many months and is not prohibitively expensive. From a performance point of view it is doubtful if you would notice a significance in changing from helium to hydrogen unless you are carrying out very precise peak parameter measurements. You would, however, notice the deterioration on changing to nitrogen.

If you are faced with a marginal separation, the solution is more likely to lie in an increase in column length, more theoretical plates, or a change to a different stationary phase rather than a change of carrier gas.

Helium is the carrier gas of choice if you are going to interface to a mass spectrometer.

Self-indicating traps to scrub the gas of oxygen and moisture are available from chromatography suppliers and should be installed in the carrier gas line to the instrument.

Do not turn off carrier gases at night or over the weekends, it is not worth the saving in gas, even if the instrument power is off.

Measuring the flow rate from a capillary column is quite possible using a small-volume soap bubble flow meter. Make your measurement at room temperature as the column flow rate reduces with temperature.

This raises an important issue. It was observed that if a chromatographic elution was initiated at 70 °C with a flow rate of 2 ml/min then by the time the temperature had been programmed to 300 °C the apparent flow rate had dropped to less than 1 ml/min.

The argument then followed that the latter part of the elution was being carried out under non-optimum flow rate conditions. In response, instrument companies developed constant flow devices based on inlet pressure programming, to add on to chromatographs. A significant analysis time saving resulted, up to 50% on occasions which in the case of long runs of analyses provided some justification for the expenditure on the equipment. There was also a significant reduction in apparent column bleed with pressure programming and this was particularly noticeable in HRGC–MS.

However, this solution has been challenged by pointing out that as the temperature in the column increased, the gas volume expanded such that the volume flow rate through the column remained essentially constant and near the optimum.

From my point of view the analysis time saving was worth having, irrespective of the interesting theoretical debate.

4.7. OPTIMISATION OF SEPARATIONS

In our discussions in Part 3 we considered the manner in which the attractive forces between solutes and stationary phases were dependent upon the chemical structures of both and the way in which they interacted with each other to effect the chromatographic separation.

In the packed column environment, the resolving power in terms of column efficiency, i.e. theoretical plates, was limited, typically 3000 plates, depending on how well your column was prepared. Therefore, choice of stationary phase was important since the selectivity of the column was a dominant influence on separation.

To a much greater extent, efficiency dominates capillary column resolution since 100 000 theoretical plates is the order of the day. The rules of selectivity do not change and are still appropriate in optimising separations. However, it is much more likely that a general purpose column, e.g. DB-5, will be an adequate work-horse for a large proportion of your separations.

Retention time and resolution often go hand-in-hand and frequently we find that we reduce the column operating temperature or lengthen the column, all with the objective of increasing retention in the hope of achieving resolution of a difficult peak pair.

If we are to be successful in optimisation it is important that we consider the variables in a logical manner. In capillary chromatography we have a simple relationship which includes all the variables of relevance to our activity. The following relationship can be derived:

$$t'_R = \frac{C_S}{C_M} \times \frac{2d_f}{r} \times \frac{L}{\bar{u}}$$

where t'_R adjusted retention time
$C_S/C_M = K_d$ distribution constant
d_f film thickness
r the radius of the column
L the length of the column
$\bar{u}$ the average linear velocity of the mobile phase.

Note that $t'_R = t_R - t_M$

where t_M gas hold-up time; time required for the retention of a non-retained peak
and t_R retention time; the time from the point of injection to the peak maximum.

SAQ 4.7a	The actual time spent by all solute molecules in the gas phase as they pass through the column is the same. True or False?

SAQ 4.7a

Let us briefly consider each of the terms of the equation in turn.

t'_R As outlined above this is the adjusted retention time, the end result which we are endeavouring to influence.

C_S/C_M This is the distribution constant K_d, the key factor in chromatography. You will be well aware by now that the retention characteristics of the solute molecule are determined by their distribution between the gas and stationary phases. The greater the concentration in the stationary phase and the lower the concentration in the gas phase, the longer the retention time. You may find it easier to visualise this concept if we change this statement to 'the greater the affinity of the solute molecule for the stationary phase, the greater the retention', which means the same.

One fundamental which may clarify your thinking in chromatography is the fact that all solute molecules passing through a column spend the *same* amount of time in the mobile phase. A fact worth thinking about!

The distribution constant may be changed by changing the temperature. In a temperature programmed chromatogram we are, in effect, programming the distribution constant.

Alternatively to achieve a more significant change in K_d, we may change the stationary or bonded phase.

d_f Film thickness influences chromatographic performance in two ways. If the film thickness is increased, retention times will be increased since there is a potentially greater mass of phase for the solute molecules to transit during their time in the phase film. In addition, thicker films provide greater sample handling capacity. This will not influence retention but it can influence peak shape and prompt us to optimise the method to avoid peak overloading which could give rise to unsatisfactory quantitative results.

r   Increasing the radius of the column will reduce retention times, since the solute molecules will spend more time in the gas phase between stationary phase contacts. For a separation which is readily achievable on a 0.32 mm i.d. column there will be a significant analysis time-saving to be gained in transferring the method to a 0.5 mm i.d. 'Megabore' column. If more retention is required then progress to a 0.25 mm or a 0.1 mm i.d. column.

L   Column length and efficiency go hand-in-hand, provided of course you are working near optimum conditions. Be careful about column efficiency. A longer column will give you longer retention times and more efficiency but resolution only increases in proportion to the square root of the efficiency since:

$$R_S = \frac{\sqrt{N}}{4}\left(\frac{\alpha - 1}{\alpha}\right)\left(\frac{K}{K + 1}\right)$$

If you have doubts about this then refer to *ACOL: Chromatographic Separations, Part 3*.

∏ Now try these calculations:

A 15 m column gives 40 000 plates and a 30 m column gives 80 000 plates. By how much will the resolution be enhanced by transferring the analysis to the longer column?

If your answer came out at 1.4 then that is correct. The resolution enhancement is proportional to the ratio of the square roots of the efficiencies, thus:

$$\sqrt{80\,000} / \sqrt{40\,000} = 1.4$$

Efficiency is inversely proportional to column diameter. The 15 m × 0.32 mm i.d. column has an efficiency of 40 000 plates. If we reduce the internal diameter to 0.25 mm i.d., will the efficiency increase or decrease and by how much?

You should have come to the conclusion that the efficiency will increase. The increase in the number of theoretical plates will be very significant as shown by the calculation below.

$$40\,000 \times (0.32/0.25) = 51\,200 \text{ plates}$$

Now double the length of this column and calculate the resolution enhancement on our original 40 000 plate column.

Your answer should be 1.6 as shown below.

$$\sqrt{102\,000} / \sqrt{40\,000} = 1.6$$

$\bar{u}$ The effect of average linear gas velocity, or in effect the flow-rate we set for our column is straightforward. Increase the flow-rate and you shorten the analysis; reduce the flow-rate and you extend the analysis time. Remember, however, that a large change in flow rate may take you out of the region of optimum flow, with a marked reduction in efficiency and resolution.

SAQ 4.7b   With reference to the relationship below, predict how the retention time will change on changing the parameters.

$$t'_R = \frac{C_S}{C_M} \times \frac{2d_f}{r} \times \frac{L}{\bar{u}}$$

(1) Decreasing the film thickness, d_f.
(2) Increasing the internal diameter.
(3) Reducing the linear gas velocity, $\bar{u}$.
(4) Increasing the column length, L.
(5) Increasing the temperature.

SAQ 4.7b

We will not dwell on the appropriateness of particular phases for particular separations.

Solutes whose boiling points differ, if only slightly, will almost certainly be resolved on the non-polar methyl silicone phase columns, irrespective of their functionality.

Phenyl methyl silicones can share electrons with aromatic compounds and thus enhance selectivity to this class of solute.

The cyanopropyl-containing phases and the polyethylene glycol based phases exhibit strong dipole moments and strong hydrogen bonding characteristics suitable for the resolution of polar solutes.

Most catalogues provide a good selection of typical separations to illustrate the application of the available phases and column dimensions. You will also find reference to a number of specialised columns which are used in regulatory testing, particularly in the environmental analysis field.

In an ideal world it would be most convenient if we could carry out all our analysis under the same conditions. Gas chromatography is not an ideal world, but I have found that by having a standard set of analysis conditions I can frequently resolve problems associated with chemical production activity very quickly and be able to switch from one process set of samples to another without the risk of setting in the wrong analysis parameters.

Having a standard condition set may represent a degree of analytical overkill with more component resolution than is necessary and excessive analysis time after the elution of the last eluted peak. Certainly in initial or investigative analysis, a degree of extended analysis time is a wise precaution since you need to be sure that all the possible components have been eluted from the column before you begin the next analysis. Indeed the very late eluting components may contain vital information to your understanding of the production problem.

Once you are sure of your separation then you can optimise it in terms of temperatures and time.

The following conditions are a suggestion as a starting point for HRGC analysis:

Column:	J & W DB-5, 30 m × 0.25 mm i.d., 0.25 μm film
Temperatures:	70 °C isothermal for 2 min programme at 16 °C to 300 °C, 10 min final isothermal period
Carrier gas:	helium at 2 ml/min
Sample solvent:	AnalaR ethyl acetate.

From many years of experience I have found ethyl acetate to be the ideal solvent for HRGC analysis. I have also found that if a sample dissolves in ethyl acetate it will almost certainly run on a capillary column. If it proves to be difficult to dissolve then you may have to push the analytical conditions pretty hard. If the solute does not dissolve in ethyl acetate then it probably will not run on HRGC and you may have to resort to HPLC or SFC. I call this the 'Fowlis Rule'.

At this stage we shall not discuss sample loading or solution concentration since this is very much injection-system dependent and will be considered in Part 5.

SAQ 4.7c	Suggest a column and operating conditions which will provide you with a good chance of successfully analysing more than 70% of your samples at the first attempt.

SAQ 4.7d The phantom chromatographer has struck during the night and stolen your favourite DB-1 column. This was a 30 m × 0.25 mm i.d. column with an efficiency at optimum gas velocity of 112 000 theoretical plates. Urgent analysis is required and the only DB-1 column you have available is 15 m × 0.32 mm i.d. Calculate by how much your resolution is likely to be reduced by using this column. Suggest an alternative column for your analysis.

4.8. COLUMN MAINTENANCE

Maintenance is an essential part of capillary column operation. It is quite true that bonded phase capillary columns will give several years continuous service but this will only be the case if they are properly maintained.

Having installed your column, it is valuable to determine the baseline performance so that it is then easier to assess whether or not the column performance is satisfactory at a later date.

The manufacturer will have supplied you with a specimen chromatogram run on your column and will probably also have supplied you with a vial containing the test mixture. It is a good idea to run the mixture with the column installed in your chromatograph and, if the results are comparable, you can be sure that you have carried out the installation correctly.

If your chromatogram does not compare with that supplied do not panic. Start again, logically working your way through the installation, making sure at each stage you have everything in order. It will all come right.

Having run your test chromatogram successfully, file it away carefully together with the manufacturer's chromatogram.

If you are using the column a lot then it is prudent to run your test mixture frequently, maybe weekly or bi-weekly, it is for you to decide. Examine your chromatograms carefully, looking for any loss of resolution or any indication of peak tailing. If performance has deteriorated then the problem almost certainly lies in the first few inches of the column.

To some extent the method of injection may also effect performance but consideration of this influence will be left until Part 5.

Carefully detach the column at the injector end and, using your cutter, remove approximately 6 in (15 cm) of column. Do not throw it away because you will want to examine it under the microscope. Reconnect the column, restore the carrier gas and leave it for 45 min before running up the heaters.

Run your test mixture again and you will notice a significant improvement in your column performance.

Examine your detached length under a microscope. The polyimide coating is transparent and you will observe irregular fragments and particles inside the column. These are the adsorptive site materials which were adversely degrading your chromatography.

Removing 6 in may not in some cases be sufficient, as I have found particulate material 18 in into the column. If you are using a retention gap it is retention gap material that you remove, not the analytical column. Particles sometimes end up within the press-fit connectors so it is always worth checking there using a magnifying glass.

Remember that not all components respond to adsorptive sites in the same way, so do not expect all peaks to show tailing.

In very severe cases of deterioration it may be necessary to back-wash the column with a series of solvents, but really this is a desperate measure and should only be considered after acquiring the necessary equipment and instructions.

When not in use, protect your column from air by pressing the ends into a silicone septum and storing it is its box in a suitable storage cupboard.

Learning Objectives of Part 4

After studying the material in Part 4, you should be able to:

● describe the construction of capillary columns;

● discuss the range of capillary columns, their phase characteristics and their application;

● successfully install a capillary column in a gas chromatograph;

● discuss the advantages and disadvantages of different carrier gases;

● recognise the symptoms associated with inadequate fitting, sample residues and column degradation;

● establish conditions for general use;

● understand and apply the relationship

$$t'_{R} = \frac{C_{S}}{C_{M}} \times \frac{2d_{f}}{r} \times \frac{L}{\bar{u}}$$

5. Injection Systems for High Resolution Gas Chromatography (HRGC)

5.1. INTRODUCTION

Injection of samples into packed column gas chromatographs is relatively straightforward and provided the necessary skills are mastered the chromatographer can expect to achieve repeatable results.

In HRGC the capacity of the capillary columns is much less than that of the packed column and therefore much smaller loadings of solutes must be placed on the column if overloading and the associated degradation of resolution is to be avoided.

Since, in the early days of chromatography, it was not possible to manufacture a syringe with a needle of sufficiently small diameter to pass into a capillary column, vaporising injection systems were designed. These were the forerunners of the present day split/splitless injectors.

Only later, with the introduction of the fused silica syringe needle, did injection directly onto the column become a reality. However, with on-column injection, as the sample is being injected onto the column the mass of solutes injected must be lower than that used with vaporising injection where only a small portion of the sample injected actually enters the column.

In split/splitless injection, sample introduction is made using a standard microlitre syringe through a septum into the vaporising chamber. Automation is therefore no more difficult than for a packed column injection. Automation of on-column injection requires a precision engineered injection system and the use of retention gaps, but more of that later. Nevertheless, on-column injection can produce excellent quantitative analysis and in the majority of analyses is the method of choice for the discriminating chromatographer.

It is impossible to cover the topic of injection methods for capillary chromatography comprehensively in this book. We shall, however, focus on the more important aspects of the injection methods. Two books written by Dr K. Grob are recommended as a source of further reading and practically orientated information if you wish to achieve an in-depth understanding of the subject.

Classical Split and Splitless Injection in Capillary Gas Chromatography and *On-column Injection in Capillary Gas Chromatography* are listed in the bibliography and are written for easy learning with key material highlighted.

SAQ 5.1a Which of the following statements regarding injection into capillary columns are correct? Circle YES or NO.

(a) Normal GC syringes cannot be used against the high pressures used in HRGC. YES/NO

(b) HRGC columns have limited capacity requiring lower sample loading. YES/NO

(c) Conventional syringes will not fit into a capillary column. YES/NO

(d) Samples must be completely vaporised before they enter the capillary column. YES/NO

(e) The two most common injection techniques in HRGC are split/splitless vaporisation and cold on-column injection. YES/NO

(f) All injectors are non-discriminating and give good quantitative results in HRGC. YES/NO

SAQ 5.1a

5.2. THE SPLIT/SPLITLESS INJECTOR

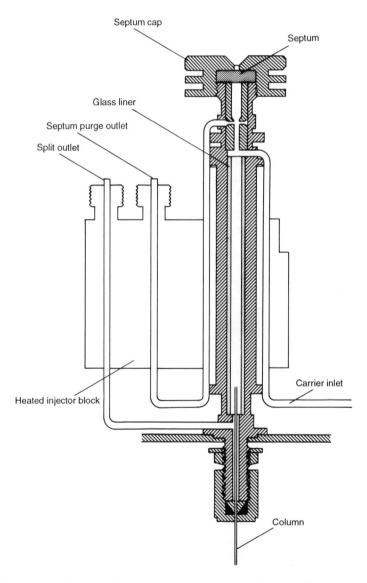

Fig. 5.2a. *Split–splitless vaporising injector for capillary columns*

Figure 5.2a shows the construction of a typical split/splitless injector. As the name suggests, the injector may be operated in either of two distinct modes depending upon the analysis to be carried out.

The injector comprises a heated chamber containing a glass liner into which the sample is injected through an injection septum. The chamber is heated independently of the chromatographic oven with the temperature normally set slightly in excess of the final elution temperature of the chromatogram.

The carrier gas enters the chamber and can leave by three different routes when used in the split mode. The injected sample vaporises rapidly to form a mixture of carrier gas, solvent vapour and vaporised solutes. A portion of this vapour mixture passes onto the column but the greater volume leaves through the split valve exit. The ratio of the split flow to the column flow rate is called the split ratio. Ratios of 50:1 and 100:1 are fairly common. A small septum purge flow prevents septum bleed components entering the column system. The split flow and the septum purge flows on the injector are controlled by needle valves.

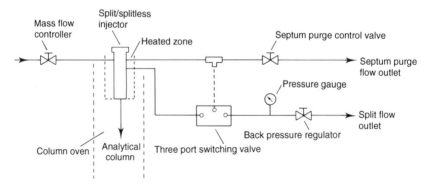

Fig. 5.2b. *Flow diagram showing the use of a split/splitless injector operating in split mode. Due to short sample residence time inside the inlet, the technique requires rapid volatilisation; thus, inlet temperature must be high enough to ensure this. The back pressure regulator in the split vent path maintains constant pressure at the head of the column. Total inlet flow, controlled by a mass flow controller, divides between a septum purge path and flow down the inlet insert. Flow through the insert is divided again, between flow into the column and flow around the bottom of the insert, up between the outside of the insert and inlet body, and onto the split vent*

In other designs the split flow is controlled by a back pressure regulator and the volume flow by a mass flow controller, Fig. 5.2b.

Thus if the column flow rate is set at 1.0 ml/min and the flow from the split control valve set at 50 ml/min then the split ratio is 50:1.

SAQ 5.2a | There are three ways in which gas may leave a vaporising injector; what are they and what is their function?

The position of the needle tip at injection relative to the end of the capillary column within the vaporising chamber is important if optimum performance is to be realised.

Care should also be taken in vaporising injection methods to ensure that the volume of vapour generated does not exceed the capacity of the injection chamber. Manufacturers usually provide adequate guidelines. Excess vapour can, in some cases, find its way into other parts of the system plumbing where solutes condense only to reappear later as ghost peaks and baseline wanderings.

Discrimination, i.e. the production of a chromatogram which is not truly representative of the actual composition of the mixture, is an accepted deficiency of split/splitless injectors and while some are worse than others, failure to set up the device properly or to use the specified syringe will certainly make matters worse.

Discrimination is a function of the relative rates of volatilisation of the solutes in the chamber, the composition of the solvent, carrier gas, solute vapour mixture, the flow and mixing characteristics of the chamber and the dwell time in the chamber. Retention of less volatile material within the syringe needle and the possibility of sample droplets falling directly onto the column entrance further complicate the injection environment.

∏ Discrimination will give rise to poor quantitative analysis. Can you suggest another factor which might also cause problems in quantitative analysis?

You may have suggested gas leaks. I once had difficulty with one instrument which was fitted with a split/splitless injector and for a long time I put down the low sensitivity and the poor quantitative results to poor equipment design. I subsequently discovered that there was a significant leak on the connection to the split flow control valve and once this was rectified, sensitivity improved and data became acceptable. The actual split flow had been very much greater than I had been measuring.

∏ How could I have avoided this problem?

Obviously I should have checked the system for leaks. In the case of the injector, if I had closed the split flow and septum purge valves and then turned off the gas supply to the injector the inlet pressure should have remained essentially constant. The flow through the column only reduces the pressure very slowly. If there was any significant leak then the pressure would drop more quickly.

Build up of non-volatile residues on the glass liner can also give rise to additional decomposition and adsorption problems, and the fact that the chamber is very hot may cause decomposition or degradation of thermally labile compounds. Some workers employ a very rigorous cleaning and deactivation routine in preparing injection liners for future use. This routine is summarised below:

Place the liner in a glass ampoule or similar vessel, Fig. 5.2c.

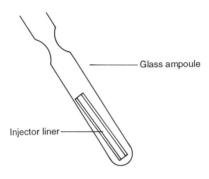

Fig. 5.2c. *Glass ampoule containing injection liner*

(a) Wash with (i)   25% nitric or hydrochloric acid, and
              (ii)   purified water, i.e. HPLC grade.

(b) Dry at 150 °C.

(c) Flush with dry nitrogen.

(d) Add hexamethyldisilazane (HMDS) or diphenyltetramethyl-disilazane (DPTMDS): two drops neat, flame seal the ampoule and heat to 140 °C and then slowly to 170 °C and then up to 300 °C. Hold at 300 °C for a minimum of 3 h.

When required for use, crack open in a fume-hood and wash with dichloromethane.

Handle with clean tweezers.

SAQ 5.2b	List three ways in which the use of a split/splitless injector may result in the resulting chromatogram not being representative of the injected sample.

SAQ 5.2b

The objective at injection is to place a narrow band of solutes onto the column. In the absence of any on-column solute focusing, the dwell time of the vaporised sample in the injection chamber is going to influence initial solute band-width and the potential resolution of the column. Thus higher split flows will result in a more rapid removal of sample material from the injector, narrower band-width and potentially sharper but smaller peaks.

Where temperature programmed analysis is being carried out, the initial bandwidth may be an insignificant problem since the volatilised components may experience refocusing once they reach the cooler column and are subject to the retentive forces of the stationary phase or solution on condensed solvent if the oven temperature has been set significantly below the boiling point of the solvent.

Low split ratio can provide higher sensitivity since more material will enter the column, provided of course that the advantage is not lost by band broadening on the column. The logical extension of this is to operate in the splitless mode. This will be covered later.

Various designs of injection liner have been developed over the years in attempts to improve the quantitative performance of vaporising injectors. Some of these are illustrated in Fig. 5.2d. It is my personal view that if a simple unpacked liner will suffice, so much the better.

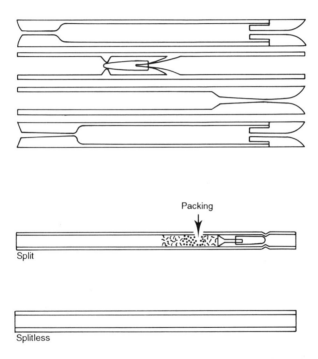

Fig. 5.2d. *A selection of the various designs of split injection liners that are currently available*

Splitless injection onto capillary columns, where the split valve is closed during the injection period, achieves higher sensitivity than the split mode since, potentially, there should be no loss of sample material through the split outlet. If you consider the volume of the injector and the flow rate onto the column, then it is obvious that passage of the total evaporated sample onto the column is going to take a considerable time

— tens of seconds. Unless solute focussing effects are active, this may result in a very wide initial on-column solute band. As a general rule, if the initial temperature of the column is 80 degrees below the elution temperature of the first component of interest, then that component will become focussed as it enters the analytical column from the injector. Focussing may be due to both partitioning effects on the column or to focussing in recondensed solvent in the column inlet.

In practice, the split valve is normally opened after a predetermined period to remove residual traces of vapours which might otherwise continue to pass onto the column during the early stages of the analysis.

Headspace samples may also be analysed using the vaporising injector in splitless mode, as all the vapour sample may be passed onto the column and, provided solute focussing takes place, reasonable sensitivity can be achieved.

An alternative to the classical split/splitless injector is the programmed temperature vaporising injector sometimes called the split/splitless injector.

Figure 5.2e shows a simplified diagram of this device. In most respects it is similar to the normal vaporising injector except that the dimensions are smaller and the thermal mass is much reduced. The body of the injector can be rapidly cooled and heated and the smaller internal diameter allows rapid and uniform heating of the vaporising chamber.

The vaporising chamber glass liner is normally packed with quartz wool in order to prevent droplets reaching the column. This packing material also acts as a support from which the solvent and solutes are vaporised.

The initial injection temperature is normally set at about 10 °C below the boiling point of the sample solvent. A few seconds after removal of the syringe needle the vaporiser is heated to a predetermined temperature, typically 300 °C in 20–30 s.

Since evaporation of the solvent is likely to take place first in the

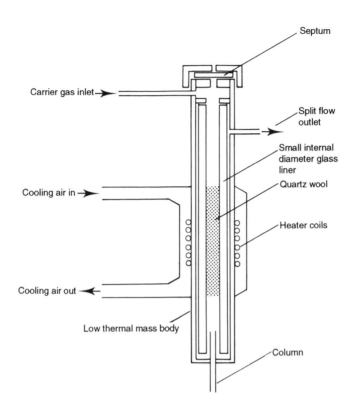

Fig. 5.2e. *Programmed temperature vaporising injector*

initial temperature, it is feasible to vent most of the solvent through the split valve and then later close the valve to pass the vaporising solutes onto the analytical column.

In most other respects, the programmed temperature vaporising injector performs as other vaporising types. Some say that discrimination problems are less and certainly the thermal shock to the sample should be less severe which may improve quantitative accuracy.

From the foregoing discussions, you should now have realised that vaporising injectors are not noted for their quantitative performance, although with care and attention to detail, acceptable data can be obtained.

5.3. THE COLD ON-COLUMN INJECTOR

In our earlier discussion of injection into packed columns I expressed the opinion that injection of the sample or sample solution directly onto the column, allowing equilibrium to be established prior to commencing elution was the ideal manner in which to carry out chromatographic analysis.

This opinion also holds for capillary columns, but until fairly recently direct injection into capillary columns was not technically possible.

The introduction of on-column injectors and syringes with fused silica needles revolutionised HRGC for me. At last I could inject my samples directly into the column.

There are a number of differing designs of on-column injector, some of which can be fitted as extras by the chromatographer. Figure 5.3a shows the construction of a particularly good on-cold injection valve. Unlike normal septum injectors, this device has no septum and syringe injection is made through a precisely engineered rotating valve with the syringe needle tip reaching into the column oven. The end of the column is aligned by the column ferrule and the fixing nut.

The body of this injector is cooled by fan-driven air taken from the back of the instrument and it remains cool even when the column oven has been programmed to high temperatures. This is called primary cooling. A novel addition is secondary cooling, where cool air is allowed to flow around the injection area of the column during the injection period.

∏ What do you consider is the merit in these additional cooling systems?

These cooling systems eliminate evaporation of sample and solvent from the syringe needle prior to injection and assist the focussing of the solute band at the beginning of the column. They also serve to protect thermally sensitive analytes until they have equilibrated in the inert environment.

Provided the column has been correctly aligned with the needle port,

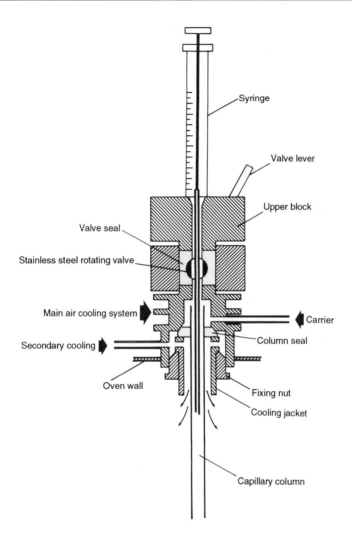

Fig. 5.3a. *CEST cold on-column injector featuring a patented secondary cooling system*

and this can be achieved by setting up the column with the syringe in place in the injection valve, this on-column injector is very simple to use and excellent repeatability of injection can be obtained. It is quite possible to achieve a percentage relative standard deviation *(RSD)* on peak areas from sequential injections of 0.5. Retention time repeatability can also be excellent.

SAQ 5.3a | List three advantages of cold on-column injection.

The use of a fused silica needle syringe is no different from that of any other syringe. You will probably find that it is more robust and lasts longer than its metal needle cousins.

SAQ 5.3b | May both fused silica and metal needle syringes be used in on-column injection?

SAQ 5.3b

Apart from the usual precautions of carefully wiping the outside of the syringe needle to remove adhering solution, the only recommended difference in operation is to withdraw the sample away from the syringe needle tip a few centimetres. This you can do visually as you can see the liquid within the needle. The reason for doing this is as follows.

The fused silica needle is a close fit inside the column and if injection is made slowly there is a likelihood of the liquid being attracted to the capillary space between the column wall and the fused silica needle, Fig. 5.3b. If this occurs, sample discrimination is likely and sample material may be withdrawn on the syringe needle, resulting in poor repeatability and contamination of the injection valve.

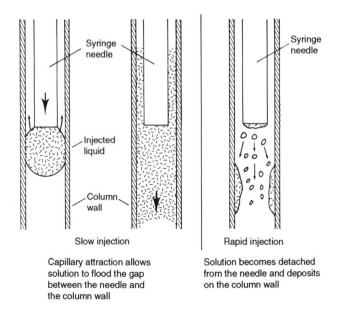

Slow injection

Capillary attraction allows solution to flood the gap between the needle and the column wall

Rapid injection

Solution becomes detached from the needle and deposits on the column wall

Fig. 5.3b *Effect of different rates of injection into the column*

This potential problem can be avoided if the plunger is depressed sufficiently quickly to allow the column of liquid to break as it leaves the needle tip. Having withdrawn the liquid part way into the needle allows the solution plug to accelerate before it starts to leave the syringe tip, further eliminating the possibility of syringe/wall solution attraction.

On-column injection is normally carried out at or near the boiling point of the solvent. A temperature five degrees lower than the boiling point will promote the possibility of solute focussing by the solvent effect, i.e. the solvent is acting as a stationary phase.

SAQ 5.3c   Why is it important to depress the syringe plunger smartly at the moment of injection?

Some chromatographers have expressed concern that carrier gas passes around the syringe needle during the injection and that this may result in a loss of sample. In practice this does not appear to be a problem since the point of injection is well below the carrier gas inlet to the column and there is always sufficient pressure, even with the injection valve open to maintain a positive carrier gas flow through the column. Indeed, as I shall describe later, even in the case of the transfer of headspace vapours, sample loss does not appear to be significant.

Should peak shape and column performance begin to degrade, then almost certainly this is due to a build up of residues in the first few centimetres of the column. These residues may affect some components more than others and in severe cases may cause decomposition of sensitive components.

The situation can normally be rectified by disconnecting the column from the injectors, cutting off 5–10 cm from the injection end of the column and reconnecting to the injector. Remember to allow time for the air which will have entered the column to be displaced by the carrier gas before heating up the column.

∏ Cutting off lengths of column will reduce the resolving power, true or false?

The answer is obviously 'true' but the question is really by how much. Remember from our earlier discussions in Section 4.7 that resolution was a function of the square root of the number of theoretical plates. 10 cm taken off a 30 m column showing 100 000 theoretical plates is only a loss of 333 plates so:

$$\sqrt{99\,667} \,/\, \sqrt{100\,000} = 0.998$$

A reduction of 0.002 could not be detected.

Cold on-column injection is undoubtedly the method of choice for quantitative HRGC analysis and particularly for the analysis of thermally labile or sensitive materials.

5.4. RETENTION GAPS

Retention gaps are lengths of deactivated fused silica tube which have no bonded stationary phase attached. They serve a number of purposes but were primarily introduced to enable the wider metal syringe needles on automatic on-column injectors to be used with narrow bore capillary columns. Retention gaps may also be used where injection of large sample volumes is required.

One end of the retention gap is connected to the injector. Normally, wider bore capillary, 0.5 mm i.d., is used and it is particularly important to ensure that the fused silica is cleanly cut with no crushed ends or loose polymide coating.

The retention gap may be connected to the analytical column by means of a suitable connector, see Section 4.5. Glass press-fit connectors are the easiest to use provided you take care to ensure that the fused silica is cleanly cut and you check carefully for leaks. As mentioned previously, gas leaks are bad news for columns.

The fact that the internal diameter of the retention gap may be greater than that of the column has no effect on the performance of the system.

At injection the sample solution spreads out over the retention gap surface. However, for the process to work effectively, the volume of solution injected must not exceed the capacity of the retention gap. It is also important that the sample solution wets the surface of the retention gap and to some extent this dictates the choice of solvent.

Methanol is generally unsuitable but many other organic solvents are satisfactory. Ethyl acetate (ethyl ethanoate) is my preferred choice, as discussed in Section 4.7.

Following injection into the retention gap the solvent starts to evaporate from the end nearest the point of injection, Fig. 5.4a. As the solvent evaporates it carries the solute molecules forward, focussing them as it does so until finally when all the solvent has evaporated, the solute molecules are further focussed by partition on reaching the stationary phase.

∏ Why do you suppose the solute molecules are not left spread out along the surface of the retention gap originally covered by the sample solution after the solvent has evaporated?

You could have suggested that since the retention gap contains no stationary phase there is greater affinity on the part of the solutes for the solvent rather than for the column wall which has no retention character and therefore as the solvent evaporates, the concentration of solutes increases in the reducing liquid volume.

This focussing process is extremely effective and even in the case of normal small-volume injections can result in a narrowing of the initial solute band prior to elution, with a consequent improvement in resolution and efficiency.

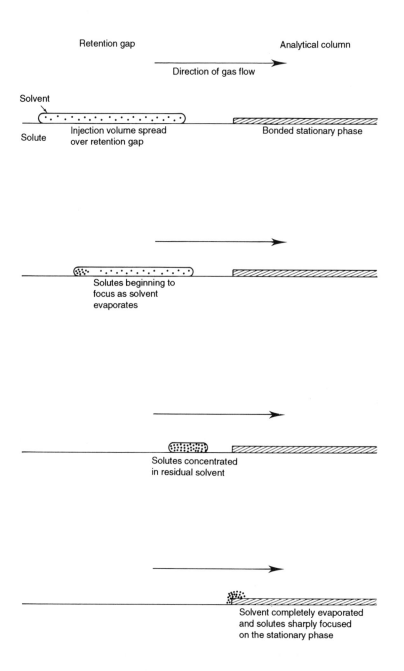

Fig. 5.4a *Solute focusing*

SAQ 5.4a | What is the function of a retention gap?

5.5. AUTOMATIC INJECTION IN HRGC

The capital cost of advanced analytical equipment can only be recovered if maximum utilisation is achieved. Automatic injection systems can provide 24 hour, 7 day a week, operation if required and, in addition, provide more repeatable and precise analysis that can be achieved on a manual basis.

Automation combined with data handling and laboratory information management systems (LIMS) leads to the automated laboratory.

As discussed earlier, injection into split/splitless vaporising injectors is no more difficult than into packed column systems and is usually based on an automated conventional microlitre syringe. Some valve switching will be necessary in splitless injection but this can easily be arranged using built-in switching signals.

Some on-column injectors also use normal syringe autosamplers with septum injection or a duckbill septum injection port. Where on-column injection requires the use of a relatively long and fine needle to reach into the column oven, then the syringe needle may not be sufficiently robust to stand up to repeated penetration of septum sealed vials and the injection septum.

An alternative approach based on the cold on-column injection valve described in Section 5.3 is shown in Fig. 5.5a. Sample solution is displaced by gas pressure from a sealed sample vial through a flow-through syringe, the syringe needle and the flow diverter assembly. After flushing for a predetermined time the liquid is locked into the syringe by valve closure, the injection valve opens and the syringe moves down onto the valve and the needle into the retention gap. At the moment of injection the syringe body moves down the needle discharging the desired volume of the solution into the retention gap. The syringe is withdrawn, the injection valve closed and the analysis commenced. Provided care is taken in aligning the autojector, the injection valve and the column, this system is excellent.

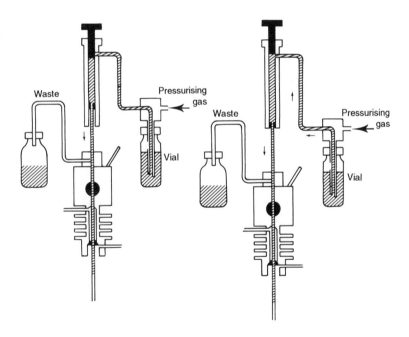

Fig. 5.5a. *Automatic on-column injection of small (up to 3 µl) and of large sample volumes by the autosampler of Carlo Erba. Small sample volumes (left) are injected by movement of the syringe barrel against the syringe needle. Injection of large volumes (right) occurs by pushing sample liquid from the pressurised sample vial through the syringe into the column inlet. The short piston at the top of the syringe starts and stops the flow of sample liquid*

5.6. LARGE VOLUME INJECTION IN HRGC

The use of retention gaps, initially to make possible automatic on-column injection as described above, also made possible large volume injection into HRGC, i.e. 10 μl to say 1 ml.

Large volume injection can provide many orders of sensitivity enhancement in determining low levels of impurities in dilute solutions. In addition, it may be used as a method of combining HPLC with HRGC, thus making it possible to combine the selectivity of HPLC with the efficiency of HRGC as a multidimensional system.

As discussed in the previous section, the most important criterion to be satisfied is that the capacity of the retention gap should not be exceeded. If by chance, or poor design, the capacity is exceeded then sample solution will reach the analytical column and establish an initial broad sample band, depending on how far the solution penetrates the column. This broad band cannot then be reconstituted by solute focussing and so resolution and efficiency will be lost.

Grob has provided guidelines for the capacity of 0.53 mm i.d. retention gaps of differing lengths and taking into consideration the column oven / solvent boiling point temperature differentials. These figures assume gas velocities in the retention gap exceeding 10–15 cm/s. Volumes are quoted in μl and are the values recommended for practical use.

Length	> 25 °C below boiling point	10–25 °C below boiling point	< 10 °C below boiling point
2 m	10	13	15
5 m	25	30	40
10 m	50	65	80
15 m	80	100	130

The use of large volume injections overcomes many of the disadvantages of solute concentration by evaporation where enhancement of solvent impurities, loss of solutes by evaporation and decomposition by heat and contamination by solvent residues all militate against low-level analysis. The large volume injection

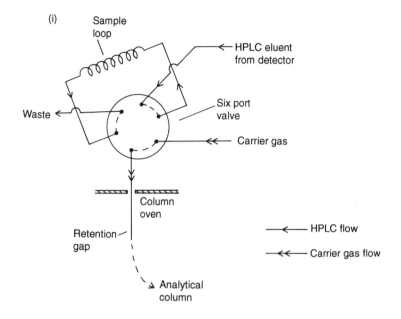

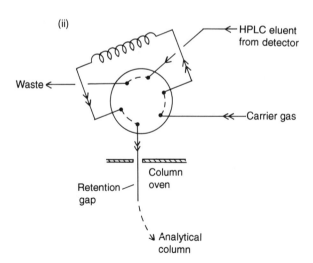

Fig. 5.6a *Method of transferring large volumes of sample solution or mobile phase fractions to the retention gap by means of a loop arrangement: (i) loading sample loop; (ii) displacing contents of sample loop into the retention gap*

technique forms the basis of the direct combination of HPLC and HRGC. Generally the combination of HPLC and HRGC should be restricted to normal-phase HPLC where the mobile phases comprise organic solvent mixtures and are free from water and salts which might degrade the column.

Large volumes of sample solution or mobile phase fractions may be transferred to the retention gap by displacement from a loop (see Fig. 5.6a), or by diversion of the flow through an automatic injector based on the cold on-column injection valve described in Section 5.3 (Fig. 5.6b). In the case of the latter, large volume transfer works on the

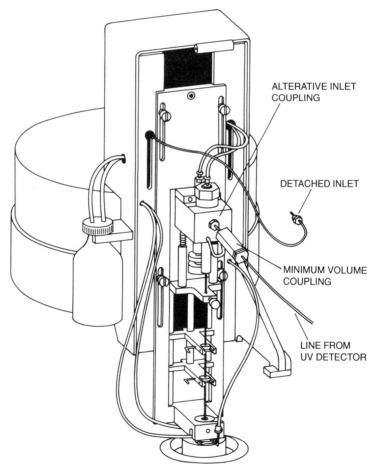

Fig. 5.6b. *Carlo Erba AS550 Autosampler head assembly*

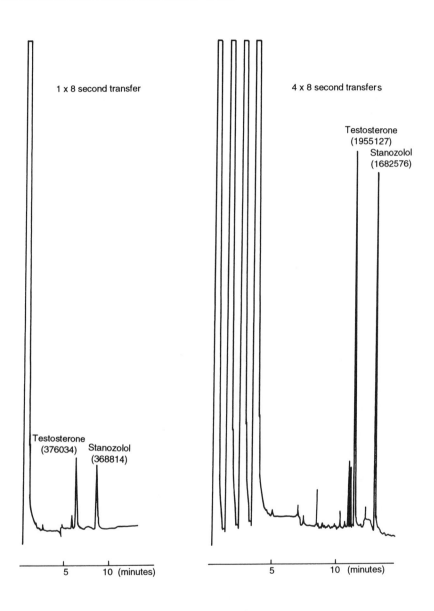

Fig. 5.6c. *Sequential large volume solution transfers. This figure illustrates how a number of large volume solution transfers may be carried out prior to elution of the focused solutes. Note that retention time becomes meaningless in multiple injection sampling (numbers in brackets signify peak area counts)*

basis of a continuous flow of solution or mobile phase through the syringe and needle and into the retention gap. The volume injected is determined by the transfer time and the flow rate.

If the conditions are optimised such that the solutes remain focused at the beginning of the analytical column following evaporation of the solvent, then it is quite feasible to make a further large volume injection to promote additional sensitivity enhancement. On one occasion I made four large volume injections prior to commencing the solute elution (see Fig. 5.6c).

Normal 4.6 mm i.d. HPLC column flow-rates are typically 1–2 ml/min and as a result, the components of interest are present in a rather larger volume of mobile phase than is ideal for HPLC–HRGC. This problem can be eased by reducing the internal diameter of the HPLC column.

Consider the LC peak eluting from a 4.6 mm i.d. column with a band-width of 1 min at a flow rate of 1.0 ml/min. The component will be eluted in 1000 μl of mobile phase. As column flow rate is directly proportional to the cross-sectional area of the column, a reduction in the internal diameter from 4.6 mm to 3.1 mm will result in reduction in the band elution volume to 450 μl. A further reduction to 2.0 mm i.d. would further reduce the band volume to 190 μl.

These calculations assume that the chromatographic peak is still eluted in one minute and that it is not necessary to reduce the sample loading on the column. It should be noted, incidentally, that reduction of column diameter and flow rate provide additional sensitivity to detection since the response of the UV detector is proportional to the concentration of solute in the liquid phase. Thus, a 200 ng solute peak eluted in 1 min at a flow rate of 1000 μl/min corresponds to an average concentration in the cell of 0.2 ng/μl. At 450 μl/min with the same solute loading the average concentration increases to 0.4 ng/μl, approximately doubling the sensitivity of the method.

This is diverging slightly but HPLC–HRGC is more successful at lower flow rates if direct transfer is used and the additional sensitivity of the HPLC part of the system comes as an added bonus.

SAQ 5.6a	What is the advantage of combining HPLC with HRGC?

Learning Objectives of Part 5

After studying the material in Part 5, you should be able to:

- describe the construction of the injection systems commonly used in capillary column gas chromatography;

- describe the principles upon which these devices are based, their relative merits and their limitations;

- successfully connect a column to an injection device;

- carry out successful analysis using the injection method discussed.

6. Detectors

6.1. INTRODUCTION

Like almost every other analytical technique, the success of gas chromatography is dependent to a large extent upon the efficiency of the detection system and its suitability for its intended purpose.

Detectors first used in gas chromatography were based on the thermal properties of the gas phase, i.e. the katharometer or hot wire detector, the gas density balance and the flame thermocouple detector. These were effective but not very sensitive and in some cases difficult to use.

Π What property of an organic compound would the flame thermocouple detect?

The answer must be the heat of combustion of the compound as it elutes from the column and burns in the flame. The flame thermocouple detector was very soon superseded by the much more sensitive flame ionisation detector.

The development of sensitive detectors for gas chromatography took place early in the development of the technique. Indeed the sensitivity of the early argon ionisation and electron capture detectors has not been improved upon in recent times even though the combination of gas chromatography with mass spectrometry and infrared spectrophotometry dates back to the early 1960s. However, modern combinations based on capillary column chromatography have great advantages in ease of operation and data handling, being PC based.

The argon ionisation detector was perhaps the first sensitive detector, but its use in quantitative analysis proved to be less than satisfactory due to its non-linear response characteristics. It was about the time that these characteristics were being thoroughly investigated by Dr R. P. W. Scott and the author that the flame ionisation detector (FID) was introduced.

Under the same evaluation conditions the flame ionisation detector proved to have a wider linear dynamic range and as a result this detector became the most universally adopted method for detection in gas chromatography, Fig. 6.1a.

The halogen sensitive electron capture detector was invented by J.E. Lovelock, also in the early 1960s, and still remains the most sensitive detection system for halogenated environmental pollutants.

∏ List the detector characteristics you consider to be important.

You could have chosen from the following:

> High sensitivity
>
> Universal response
>
> Wide linear dynamic range
>
> Fast speed of response
>
> Good stability
>
> Low background noise
>
> Reliability
>
> Ease of operation.

To an extent, some of these criteria are actually complementary, for example, high sensitivity implies fast response and low background noise. Fast peaks from capillary columns may only last a second or so; thus for accurate quantitative analysis and maximum sensitivity the detector must be capable of responding faithfully to the peak profile. Again a high background signal will limit the sensitivity of the detector and this will be manifested in an increased 'limit of detection'.

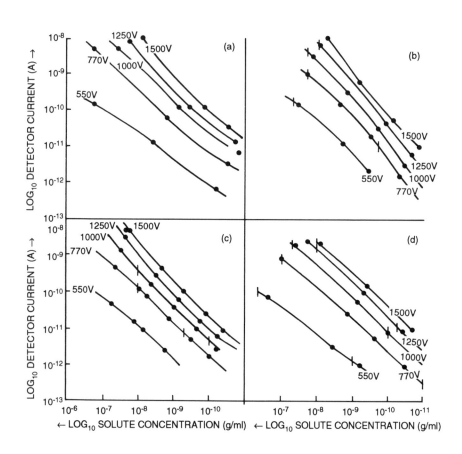

Fig. 6.1a. *Historical detector calibration data. Above, argon ionisation detector; opposite, flame ionisation detector. Solutes: (a) diisopropyl ether; (b) n-heptane; (c) toluene; (d) chlorobenzene. Notes: (i) argon ionisation detector could be operational at a number of polarising voltages, hence the series of curves; (ii) flame ionisation detector was evaluated at 50 V and 100 V, hence the use of -×- and -●- in the figures; (iii) on the axes used slope must be unity for linear response*

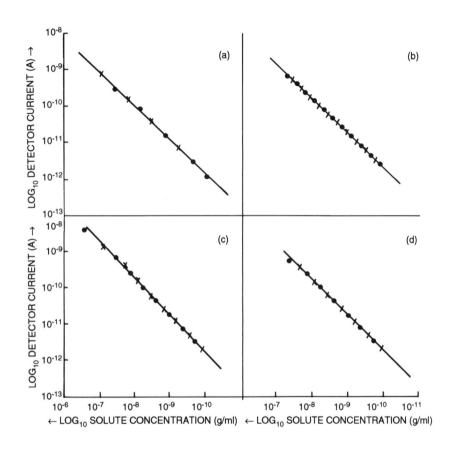

Fig. 6.1a. (*continued*)

Universal response is a 'wish-list' item, although the FID approaches this ideal to a greater extent than do other detectors.

SAQ 6.1a.

In Section 6.1 we listed the desirable characteristics of a detector as follows:

High sensitivity
Universal response
Wide linear dynamic range
Good stability
Low background noise
Reliability
Ease of operation.

Which of these characteristics do you consider to be most important in:

(a) quantitative analysis;
(b) environmental analysis;
(c) in a laboratory with a very heavy workload?

SAQ 6.1a

The following sections will discuss the construction, operation and characteristics of a number of commonly used detectors. It is important, however, to refer to the operating instructions for your own particular detector since there are often slight but significant differences in the set-up and operating conditions for individual models.

SAQ 6.1b | Which GC detector do you consider to be in most common use today?

6.2. THE FLAME IONISATION DETECTOR

The construction of a typical flame ionisation detector (FID) is shown in Fig. 6.2a, consisting essentially of a base in which the column eluent is mixed with hydrogen, a polarised jet and a cylindrical electrode arranged concentric with the flame. Air is supplied to the detector to support combustion. The assembly is contained in a stainless steel or aluminium body to which are fitted a flame ignition coil and electrical

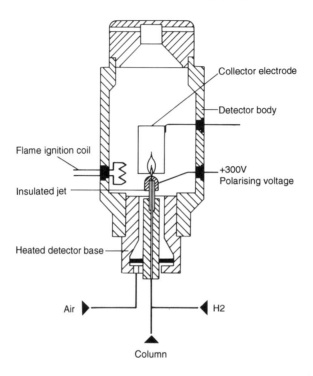

Fig. 6.2a. *The flame ionisation detector*

connections to the collecting electrode, and a polarising voltage to the detector jet. The jet itself is normally insulated electrically from the body of the detector.

Π   Why must there be a potential difference between the jet and the collector electrode in the FID?

Burning the eluting analyte in the flame generates ions but unless there is a potential difference between the jet and the collector, no ionisation current will be detected. If you know the flame is alight and that carrier gas is flowing and you think that your detector is not responding, it is a good idea to check that the polarising voltage is present.

Flame ionisation detectors are normally heated independently of the chromatographic oven. Heating is necessary in order to prevent condensation of water generated by the flame and also to prevent any hold-up of solutes as they pass from the column to the flame. It is good practice to warm up the detector before lighting the flame and to extinguish the flame by turning off the hydrogen before cooling down and switching off the instrument. A corroded detector can be very noisy and in severe cases may have to be dismantled and refurbished or even replaced.

Packed columns are normally connected to the base of the detector using a suitable coupling. Because of the relatively high flow-rates, detector geometry and dead volumes are not normally a problem. In the case of capillary columns with flow-rates of the order of 1–2 ml/min, positioning of the column end as close as possible to the flame is vitally important if optimum performance is to be realised. With the flame extinguished, the column end should be passed up through the jet and then lightly held in position by slightly tightening the coupling. Gradually draw the column end back into the detector jet until it is approximately 1–2 mm below the jet tip. Then tighten the coupling to retain it in position. Do not over tighten couplings on capillary columns.

In the presence of hydrogen and pure carrier gas, very little ionisation occurs and the base current of the detector is low. When solute molecules contained in the carrier gas elute from the column and pass into the detector they burn in the flame and in doing so, generate ions which move to the collector electrode, due to the potential difference between the jet and the electrode. The resulting ionisation current is amplified and fed to the potentiometric recorder or the data system.

The flame ionisation detector is certainly the most universal detector designed specifically for gas chromatography to date, although it is perhaps more accurate to say that it is specific for organic compound detection since compounds which do not contain carbon will not be detected on elution from the column. Thus the FID is insensitive to water but in addition the FID does not detect carbon monoxide, carbon dioxide, hydrogen cyanide, formaldehyde and formic acid.

Differences in responses between organic compounds in the flame are generally small and therefore direct comparison of one peak response with another provides a fairly accurate indication of the relative quantities of each component. Halogenated compounds are, however, an exception. The more heavily halogenated the compound, the lower the FID response. This is not often a problem since when halogenated solvents are used in analysis we are not normally interested in the magnitide of the peak relative to that of the solutes. If we were analysing say, halogenated biphenyls which are used as flame retardants in the textile industry, then to do quantitative work we would need to adopt an internal standard procedure to correct for differences in responses of the individual components. More of this later.

The wide linear dynamic range of the FID ($\sim 10^7$) makes it very suitable for accurate quantitative analysis.

∏ The FID is only part of the detection system — name the other components? Are the characteristics of these additional units important?

The amplifier and the recorder, or data system, together with the detector, make up the detection and measurement system. It is important that these units also have linear response characteristics and response rates sufficiently fast to follow accurately the fastest peak.

FIDs are generally trouble-free provided they are operated as set out in the instruction manual for the particular instrument. Dismantling and cleaning should be carried out as performance and use demands. Jets and electrodes can be cleaned and polished with very fine emery paper but be sure to wash and dry the components thoroughly before reassembly. Try to avoid getting contamination from your fingers onto the cleaned components, use gloves or tweezers. If your jet is insulated, as most are in modern instruments, check, or get one of your electronic colleagues to check that there has been no break-down of the insulation. Remember on reassembly to allow the detector to warm up before lighting the flame.

Lack of detector response usually indicates that either there is no flow through the column or the detector flame is not lit. Flame ignition can be checked by holding a cool spatula over the flame and observing condensation. If the flame is lit and there is still no response, then inject about 0.5 μl of chloroform into the column and observe the flame. Within a few minutes you should observe an intense blue flame as the solvent elutes from the column. If there is still no response on your recorder then check all the electrical connections thoroughly and ensure that all the sub-units are switched on. Lack of response is most often due to operator error!

SAQ 6.2a | List three attributes of the flame ionisation detector and one limitation.

Two variants of the FID which offer a higher degree of specificity are the thermionic or nitrogen–phosphorus detector (NPD) and the flame photometric detector (FPD).

The NPD detector, Fig. 6.2b, has an alkali metal salt bead placed in the flame, usually rubidium silicate. While this does not affect the sensitivity towards simple organic compounds, it increases the sensitivity to those compounds containing halogens, nitrogen and phosphorus by factors of up to 500.

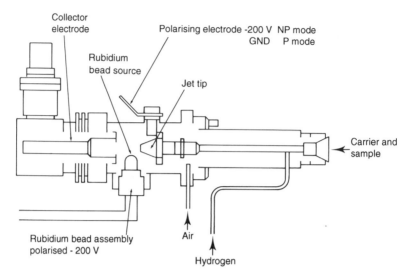

Fig. 6.2b. *The nitrogen–phosphorus detector or flame thermionic detector*

In the case of the flame photometric detector, photoemission in the flame is measured using a photomultiplier and appropriate filters rather than the ionisation current and this detector is sensitive towards sulphur and phosphorus.

In many respects the advantages of these specific detectors has been overtaken by the bench-top mass spectrometers which can be combined very effectively with capillary column gas chromatography and offer a much greater degree of sensitivity and specificity.

6.3. THE ELECTRON CAPTURE DETECTOR

The electron capture detector (ECD) was the first selective detector to be invented for gas chromatography. In many ways its origins go back to the argon ionisation detector.

The construction of the detector is shown in Fig. 6.3a. The internal chamber of the detector is kept as small as practicable and is lined with radioactive β-emitter contained is a sealed foil. The source is

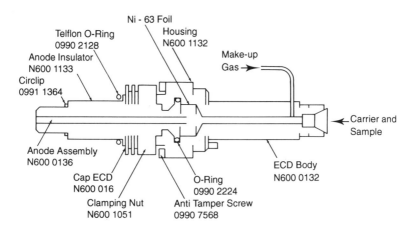

Fig. 6.3a. *The electron capture detector*

normally either ^{3}H or ^{63}Ni. The inlet tube and the cylindrical body are insulated from one another so that they may be used as the electrodes in the cell. Modern ECDs operate in a pulsed mode at 50 V amplitude and 1 μs width, and 20 to 50 μs pulse intervals.

In the ECD the carrier gas, e.g. nitrogen or argon, is ionised by the β-radiation to generate free electrons which move quickly to the anode under the influence of the potential gradient before they can recombine with the nitrogen cations to form neutral nitrogen molecules. This generates an ionisation current.

$$N_2 \underset{}{\overset{\beta}{\rightleftharpoons}} N_2^+ + e^-$$

In the presence of a component capable of capturing electrons to form an ion, then although the electron will be lost and a complementary ion formed, the total ion current will be reduced since the molecular anion will be neutralised by the nitrogen cation to yield neutral molecules.

$$AM + e^- \longrightarrow AM^-$$
$$AM^- + N_2^+ \longrightarrow AM + N_2$$

The ECD is at its most sensitive for compounds containing electronegative elements such as halogens.

Compared with other detectors the ECD is difficult to use and its quantitative performance is poor due to the limited linear dynamic range, $\times 50 - \times 100$. The secret of success appears to lie in attention to detail in terms of the quality of the carrier gas, cleanliness, maintenance of constant temperature and sample clean-up procedures.

$\prod$ In which area of chromatographic analysis is the electron capture detector most commonly used?

The electron capture detector has always been extensively used in environmental analysis for the detection and analysis of halogenated compounds down to very low levels, i.e. ppb.

6.4. THE THERMAL CONDUCTIVITY DETECTOR

The use of thermal conductivity for the monitoring of column eluents in gas chromatography goes back to the earliest days of the technique. Thermal conductivity devices had been used for some time in gas analyser systems so their use in gas chromatography was a natural development.

Figure 6.4a shows both the construction of the thermal conductivity detector (TCD) and its associated electrical circuit. The body of the detector is thermostated and small filaments are mounted in the two flow channels, one the reference channel and the other the analytical channel. Electrically the filaments form two arms of a Wheatstone bridge circuit. The circuit is balanced with pure carrier gas flowing through both flow channels of the detector. When solute molecules pass through the sample channel, the thermal conductivity of the environment is changed and the filament temperature changes with a corresponding change of resistance. This unbalances the circuit and it is this out-of-balance signal which forms the chromatographic trace.

$\prod$ Suggest what may happen if you turn off the carrier gas while the detector filaments are switched on?

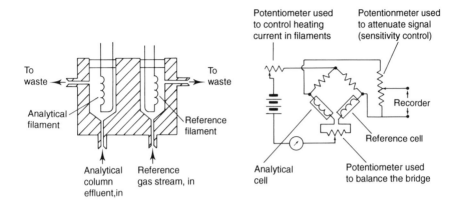

Fig. 6.4a. *Schematic diagram of katharometer and circuit*

If you think the filaments might burn out you would be correct. This is perhaps the most common mishap with thermal conductivity detectors but fortunately, apart from the embarrassment to the analyst, the filaments can be replaced in a matter of minutes, provided spares are held in stock.

Maintaining adequate stocks of spare parts and components is an important part of the chromatographers responsibilities; we should aim for 100% instrument availability.

The sensitivity of the hot wire detector is considerably less than that of the FID and its linear dynamic range is also less at 10^3. Sensitivity depends upon the difference between the thermal conductivity of the carrier gas and that of the component being detected, and it will therefore not be the same for all components in the mixture. Since most organic vapours have comparatively low thermal conductivities, sensitivity is highest when gases which have high thermal conductivities, such as hydrogen or helium, are used as carrier gases and is lowest when nitrogen or argon with lower thermal conductivities are used. The ratio of sensitivities using hydrogen, helium, nitrogen and argon as carrier gas is approximately $7:6:1:1$.

The hot wire detector (HWD) is a simple, cheap, robust and easy to use detector with limited sensitivity but is particularly suitable for gas analysis.

SAQ 6.4a	Which carrier gas would you choose for each of the following analyses, assuming that you only have a gas chromatograph fitted with a hot wire detector? (a) The determination of traces of hydrogen in air. (b) The separation of propanone and propan-2-ol.

6.5.   THE MASS SPECTROMETER

The introduction in recent years of low cost bench-top mass spectrometers which can readily be combined with high resolution gas chromatography, justifies the inclusion of the mass spectrometer as a detector in this section.

Earlier we discussed the lack of specificity of the FID and the specificity of the ECD, the NPD and the FPD detectors. The mass spectrometer can be either specific or universal as we wish depending upon how we want to operate the system and we shall discuss this in more detail later.

Mass spectrometry is based upon the ionisation of solute molecules in the ion source and the separation of the ions generated on the basis of their mass/charge ratio by an analyser unit. This may be a magnetic sector analyser, a quadrupole mass filter, or an ion trap. Ions are detected by a dynode electron multiplier. Further information may be found in *ACOL: Mass Spectrometry* and many other spectroscopy text books.

Bench-top mass spectrometers used with HRGC are normally quadrupole or ion trap instruments with electron impact and in some cases chemical ionisation sources. Figure 6.5a shows the essential building blocks of a typical capillary column gas chromatography/mass spectrometer combination.

No interface is required, apart from a heated tube between the gas chromatograph and the mass spectrometer through which the outlet end of the capillary column is fed directly into the ionisation source.

∏ Packed column gas chromatographs combined with mass spectrometers required special interface devices to allow them to work together. Why are these not necessary in HRGC–MS?

Packed column systems operated with carrier gas flow-rates of say 30–50 ml/min. The vacuum systems of the mass spectrometer could not pump away this volume flow and at the same time maintain the vacuum in the mass spectrometer. Interface devices were designed which removed the bulk of the carrier gas and allowed the solute molecules to enter the ion source. At capillary column flow-rates of 1–2 ml/min the pumping system of the mass spectrometer can maintain pressures of typically 10^{-6} torr and therefore no interface device is necessary.

Operating as a simple detector, in *acquisition mode,* the mass spectrometer scans the total mass range, typically 30–600 atomic mass units (amu), every few seconds, sums all the ions detected and then produces a trace on the control system PC screen. This is called a total ion chromatogram and is analogous to the trace we might obtain from any other detector.

∏ The mass range of most bench-top mass spectrometers is typically 30–600 atomic mass units. Is this likely to limit their use in gas chromatography?

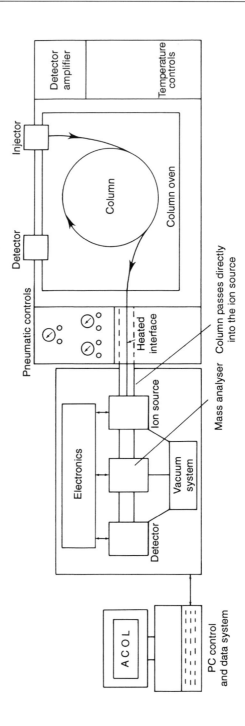

Fig. 6.5a. *HRGC/MS combination*

File: C:\CHEMPC\DATA\BSB\EXAMPLE1.D
Operator: [BSB1]
Date Acquired: 11 May 94   6:42 pm
Method File: DEFAULT.M
Sample Name:
Misc Info:
ALS vial: 1

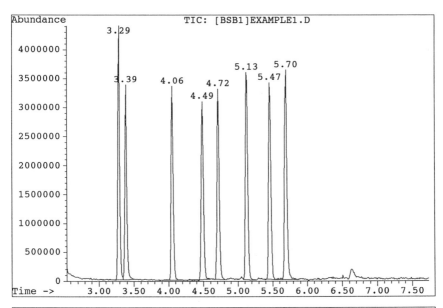

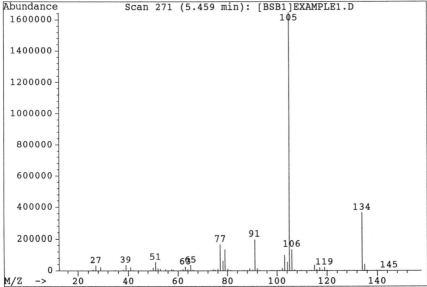

Fig. 6.5b. *Total ion chromatogram and mass spectrum of peak eluting at 5.47 min*

Probably not. The vast majority of organic compounds you will wish to analyse will fall in the mass range 100–350 amu. I think I only once broke the 500 amu barrier.

Once the chromatogram has been completed, the raw data is transferred to the hard disc storage system of the PC for subsequent manipulation in *data analysis mode.*

The total chromatogram can be presented on the screen together with the spectra of selected peaks of interest, Fig. 6.5b. Spectra can be compared with those contained in the systems spectral library to aid component identification.

In *selected ion monitoring mode,* during the acquisition, the appearance of a specific compound can be tracked by selecting an ion which is characteristic of that compound; either the molecular ion or the base peak for example. Alternatively, an ion characteristic of a group of compounds, say m/e 73 for ethyl esters, may be selected, and then these esters will be identified in the chromatogram. With most instruments a number of ions can be monitored simultaneously.

The selected ion monitoring mode increases the sensitivity of the mass spectrometer, putting it on a par with the specific detectors discussed earlier. However, it is the ability of the mass spectrometer to provide spectra of the individual components as the elute from the capillary column, which make the mass spectrometer the most universal and powerful detector for gas chromatography.

SAQ 6.5a	What are the most significant advantages of the mass spectrometer over other detectors?

SAQ 6.5a

6.6. THE INFRARED SPECTROPHOTOMETER

Development of infrared spectrophotometers as detectors for gas chromatography always lagged behind that of mass spectrometry. This was due to the inherently lower sensitivity of infrared detectors and their inability to scan the spectral range rapidly enough to provide a spectrum truly representative of the chromatographic peak. Fourier transform infrared spectrophotometers have now overcome many of the earlier limitations and such instruments are now commercially available.

The infrared spectra of the chromatographic peaks are generally recorded as vapour spectra using a long-path-length cell or multiple internal reflectance light pipe, Fig. 6.6a. The spectrum is scanned a number of times during the transit of the peak through the cell and the data processed by the spectrometer software.

There was an alternative system on the market some years ago in which the solutes eluting from the chromatographic column were

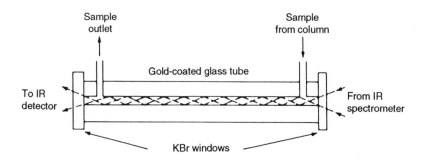

Fig. 6.6a. *Multiple internal reflectance light pipe*

condensed on a cryogenically cooled gold plated rotating drum. Reflectance spectra were then obtained when the condensed solutes moved into the light beam.

∏ What do you consider would be the major advantage of the cryofocusing method?

This method could theoretically produce a more definitive spectrum since the condensed solute would be in either the liquid or solid state. Vapour spectra lack much of the fine detail of solution, liquid film or solid-state spectra although the dissociation of hydrogen bonding in the vapour state can help in the identification of compounds such as alcohols and carboxylic acids which show broad bands due to hydrogen bonding in the liquid or solid states. You will find an excellent discussion of hydrogen bonding in infrared spectroscopy in *ACOL: Infrared Spectroscopy.*

Infrared Vapour Spectra by D. Welti is an interesting, although now slightly dated, book by one of my former colleagues, which contains a useful library and correlation of infrared vapour spectra. It also describes much of the early work in combining infrared spectrophotometry with gas chromatography.

It may be some time before infrared spectrophotometry becomes a common and widely used detection system in gas chromatography. Acceptance of a technique usually boils down to 'what do you get for your money?'

SAQ 6.6a | Classify the detectors we have considered in terms of universal response, specificity, sensitivity and ease of use, allowing 1 for excellent, 2 for good and 3 for poor.

SAQ 6.6a

Learning Objectives of Part 6

After studying the material in Part 6, you should be able to:

- discuss the principles of operation of each detector;

- describe the construction of each detector;

- discuss appropriate applications of the detectors studied;

- identify common detector faults.

7. Data Handling Systems and Quantitative Analysis

7.1. INTRODUCTION

It was not long after the introduction of commercial gas chromatographs that chromatographers, or those they reported to, wanted to know not only how many components, but how much of each component was present in sample mixtures. Since gas chromatography was able to achieve separations hitherto not possible and with considerable ease, this was not an unreasonable requirement.

Unfortunately, the non-linear response characteristics of some of the early detectors and the rather primitive integration systems, frequently resulted in questionable data and as a result the quantitative aspect of gas chromatography was not held in very great regard.

Detector calibration studies and the introduction of the flame ionisation detector brought about big improvements and these, together with faster and more precise data handling systems, repaired much of the early damage.

Problems do still occur in quantitative gas chromatography and in the remainder of this section I shall endeavour to make the reader aware of some of the pitfalls.

Data handling and quantitative analysis is comprehensively covered in *ACOL: Chromatographic Separations*, Part 4, so I shall skip the basics and concentrate on the practice. I would recommend, however, that you study Section 4.2.6, plus Section 4.3 in its entirety, of this earlier book.

SAQ 7.1a | What do you consider to be the most important characteristics of the chromatographic system if it is to produce accurate quantitative analysis?

7.2. THE CHROMATOGRAPH AND THE DATA SYSTEM

Data systems or integrators, as they were more commonly called, were perhaps the first widespread application of computers in analytical laboratories. Before the advent of the digital integrator, a number of analog integrators, based on charging an electrolytic condenser or on mechanical devices built into potentiometric recorded pen-drive mechanisms, had been developed. They were not particularly accurate but together with triangulation of the peaks and cutting and weighing the peaks, they provided the first quantitative GC analysis.

The electronic digital integrator takes the output signal from the detector amplifier, converts it into a digital signal from which it measures all the parameters necessary for data handling. Generally detectors, amplifiers and integrators (data systems) have wide linear dynamic ranges ideal for accurate quantitation.

One very important point you must have in mind is the fact that if two
peaks, which you are trying to measure, are not completely resolved,
then *any* quantitation will be an approximation and the answer will
vary depending upon how the data system handles the signal. Figure
7.2a shows a number of possibilities, none of which will be a correct
answer. This being the case, the conditions must be changed to
provide baseline separation if accurate quantitative analysis is
required.

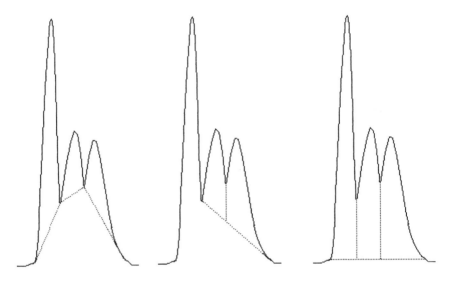

Fig. 7.2a. *Integration of unresolved peaks*

Chromatography data systems now range from basic simple single
channel integrators with appropriate built-in programmes and printer,
to laboratory information management systems (LIMS) capable of
producing complex reports, tabulations of data for trend analysis and
even the generation of Certificates of Analysis. Some data systems are
designed to control chemical processing operations based on
continuous analysis. There is a wide choice available.

SAQ 7.2a   List four possible methods of quantifying a
               chromatogram in the order you think they might
               have been introduced in GLC.

SAQ 7.2a

SAQ 7.2b   Comment on the suitability of the methods you listed in SAQ 7.2a.

SAQ 7.2c

> Under what circumstances will integration always involve a degree of approximation?

SAQ 7.2d

> What is the solution to the problem in SAQ 7.2c?

Response

7.3. INJECTION AND QUANTITATION

Apart from limited applications in process control gas chromatography and some environmental analysis, virtually all gas chromatography sampling is based on syringe injection.

Syringe design and construction has made great progress over the years. However, certainly where manual injections are made into the gas chromatograph, it is the skill of the chromatographer which is still the major influence on repeatability and precision.

SAQ 7.3a

> Why is variability of injection volume more of a problem in GC than in HPLC?

Automatic injection systems can produce remarkably repeatable results because they inject in precisely the same way time and time again. The chromatographer must endeavour to do the same.

∏ Calculate the %*RSD*, i.e. the repeatability of injection, on the following peak area measurements obtained from six repetitive injections of a sample.

This is a simple operation using a scientific calculator. First enter the peak areas and then read out the mean, $\bar{X}$, and the standard deviation, σ_{n-1}.

Then

$$\% \, \mathrm{RSD} \; = \frac{\sigma_{n-1} \times 100}{\bar{X}}$$

Now the areas:

1 214 567
1 205 349
1 224 686
1 216 001
1 205 578
1 215 856

My calculation comes to:

$$\% \, \mathrm{RSD} \; = \frac{7310 \times 100}{1\,213\,673} = 0.6\%$$

This would be a good result and should certainly be possible with a properly set up capillary system.

The level of precision we require of our injection from one chromatographic run to another depends to some extent on how we intend to handle the peak data we generate. If, for example, we are going to area normalise the data, then slight variations in the relative intensities of the successive chromatograms is not going to influence the outcome very significantly. In the situation where we are carrying out assays of a production sample against a standard where we wish to compare area response on a weight (or volume) basis accurately, then injection precision may be insufficient for us to have confidence in the significance of the result due to variation in peak areas from run to run. To overcome the injection variability we use the *internal standard method* where we add an internal standard (or internal calibrant) compound to our sample and standard mixtures and measure the ratio of the peak areas rather than the absolute area. Measuring the ratio of the compound peak area to the internal standard peak area allows us to tolerate the slight variability of injection inherent in the syringe technique. Data handling methods will be considered further in Section 7.5.

SAQ 7.3b	Which analysis and data handling method overcomes the problem of injection variability in gas chromatography?

It is important to ensure that the correct integration parameters are programmed into the data system before commencing analysis. Clearly these will depend on the equipment being used and it is very important to follow the instructions contained in the instruction manual. The most important parameters are usually those reflecting the peak-width and threshold sensitivity and the manner in which the system handles unresolved or tailing peaks.

Peak markers are usually available which indicate when peak integration starts and finishes. Sharp and symmetrical peaks present few problems; however, with tailing peaks the data system may have difficulty deciding when the peak has finished and this can result in peak area variation from run to run. Once again, as in the case of unresolved peaks, the ideal solution is to upgrade the chromatography to improve the peak shape before commencing accurate quantitative analysis.

Figure 7.3a illustrates how integration parameters may affect the integration of an asymmetric or tailing peak.

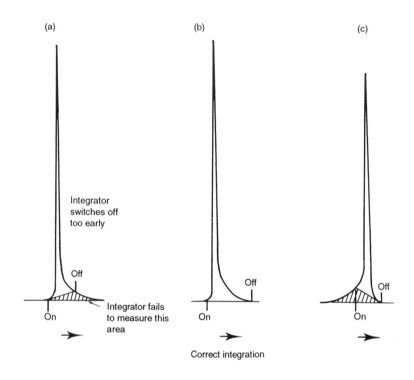

Fig. 7.3a. *Effects of different integration parameters: (a) integration switches off too early so a significant area under the peak is not integrated; (b) good integration; (c) integration starting too late — this peak is typical of thermal degradation and therefore integration will not be very meaningful*

7.4. DATA CREDIBILITY

This may seem an unlikely title for a section of this book but it has been put in for two quite different reasons.

The ease with which results can be calculated and displayed by modern systems has resulted in the chromatogram taking very much a second place. Indeed I have known of laboratories where the chromatogram was not even recorded and it was assumed that the numerical data was bound to be correct. This has led to costly mistakes from time to time.

I find many cases where the presentation of chromatograms on PC screens and from many printers is poor. Step-wise traces, blotchy printers, loss of resolution due to printer linewidth etc. It is pointless to spend several hundred pounds on a high resolution column, only to throw away considerable resolution in inferior chromatogram presentation. Indirectly, this may lead the chromatographer to place little emphasis on the trace and to assume that the numerical print-out must be correct. Here I believe we should go 'back to basics' and insist upon a quality print-out of the chromatogram since only in this way can we check that the numbers make sense. The chromatogram and the numerical data must correspond with each other.

HRGC with its very sharp and fast peaks also has its potential pit-falls. There is a temptation to run the recording system at fairly high sensitivity in order to be able to see minor impurities. Therein lies hidden danger. The major peaks have run well off-scale but do we know for certain that they are single peaks or even more importantly, can we be certain that they are not overloaded to such an extent that accurate quantitative analysis is impossible? Slow chart speeds may also conceal peak overloading so it is good practice during initial chromatograms to run at lower sensitivity and higher chart speed to ensure that any peak distortion is detected, Fig. 7.4a. You will find this problem to be more common than you expect in your HRGC analysis.

The second reason for including this section is so that you ensure that your results are credible, as well as accurate. The true analytical chemist will always go back to the source of his samples and discuss the results with the development chemist or the production chemist to ensure that they make sense. If they are not credible, then it does not matter how accurate the numbers are. It all comes down to good communications.

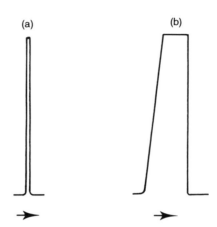

Fig. 7.4a. *Detecting overloaded peaks: (a) slow chart speed and peak well off-scale, but peak shape looks satisfactory; (b) same peak ran at high chart speed, showing clearly that the peak is overloaded*

7.5.   INTEGRATION METHODS

There are a number of alternative data handling methods which are normally programmed into even the most basic integrator or data system.

I shall restrict this discussion to the *peak area normalised method* and the internal standard method since I believe these are most widely used.

The area normalised method is the simplest routine and is the normal default method in most integrators. All the peaks in the chromatogram are integrated and then each reported as a percentage of the total of the peak areas in the complete chromatogram.

$$\% A = \frac{\text{Peak Area A} \times 100}{\Sigma(\text{Peak Area A} + \text{Peak Area B} + \ldots + \text{Peak Area } N)}$$

An integration inhibit time may be set to eliminate the solvent peak or impurities associated with the solvent. Tabulations usually include peak retention times, integrated peak areas, peak heights, an indication of how the data system handled the starting and stopping of the integration and the percentage area of the peak.

∏ The peak areas measured on the chromatogram of the methyl ester mixture shown in Fig. 1b (see Part 1) were as follows:

unknown	1025
unknown	487
unknown	2879
methyl palmitate	10 235
methyl oleate	7435
methyl linoleate	8312
methyl linolenate	2250

Note that a minimum area of 250 counts had been set into the integration routine to eliminate minor peaks from the calculation and that an integration inhibit was set to switch off after the solvent peak and related impurity peaks had eluted.

Now calculate the area normalised composition of the four methyl esters.

The results come out as follows:

methyl palmitate	31.4%
methyl oleate	22.8%
methyl linoleate	25.5%
methyl linolenate	6.9%

Here is the calculation of the methyl palmitate peak:

$$\% \text{ methyl palmitate} = \frac{10\,235 \times 100}{32\,623} = 31.4$$

The accuracy of the result depends upon the detection of all the peaks and assumes that they all have the same response.

Clearly this is not so since we have eliminated minor peaks from the integration and components do not have identical responses, but be that as it may, for many purposes the data obtained can be quite acceptable.

For many analyses, particularly those of volatile or distilled materials, the result from the method can be regarded with some confidence. In the case of wide-boiling-range mixtures and essentially unknown mixtures, then the results may be questionable since they are based only on those peaks which elute from the column. If components are retained by the column, decomposed on the column or are not detected then the data is virtually meaningless.

Be careful how you present area normalised data in your reports. The results should not be reported as assays since they are not determined with respect to a standard of known purity on a wt/wt or vol/vol basis. Furthermore, the results should be presented without units, i.e. as simply area percentages.

SAQ 7.5a	What are the limitations of the peak area normalised method?

The internal standard method overcomes many of the limitations of syringe injection as discussed earlier and also those of the area normalised procedure.

Accurately weighed sample and standard materials in solution are spiked with a suitable internal standard compound, sometimes referred to as the internal calibrant. The ratios of the sample component peak area to the internal standard peak area are calculated for both the sample chromatogram and the standard sample chromatogram and, together with the sample weight, the standard sample weight and the standard purity, are used to calculate the sample assay.

The internal standard should be chosen such that it elutes close to the compound or compounds to be analysed, not at the other end of the chromatogram.

∏ Where would you place an internal standard peak if you were to analyse the ester mixture shown in Fig. 1b by the internal standard method?

I think you would agree that it should be between the methyl palmitate peak and the methyl oleate peak. It is best if the internal standard is of similar chemical type to the analyte but this is not always possible. A range of normal alkanes up to C_{28} can be useful for this type of analysis. It is also important to check that the internal standard does not contain any impurities which may co-elute with the analyte.

As is usually the case there are a number of ways to do this. I believe the following is most practical, convenient and simplifies the calculation.

Prepare a solution of the chosen internal standard compound in the solvent to be used in the analysis. Adjust the concentration to give the required peak size. This should match the size of the anticipated analyte peak. Remember to choose a compound for the internal standard which is similar in type to the sample analyte and which has a retention time not too far away from that of the sample peak and clear of any potential impurities.

Accurately weigh sample and standard samples into suitable containers and dilute to volume with the internal standard solution. It has been my experience that modern solution dispensers provide more accurate dilution than the use of volumetric flasks.

Before commencing an extensive sequence of analyses of standard and sample solutions, it is prudent to run at least two injections of one solution to check the chromatography and repeatability of injection.

Since the same mass of internal standard is added to both sample and standard solutions, the mass of the internal standard does not come into the calculation of the assay, which is expressed as follows:

$$\% \text{ assay A} = \frac{\dfrac{\text{peak area of A in sample solution}}{\text{peak area of IS in sample solution}} \times \text{wt of standard} \times \% \text{ purity of standard}}{\dfrac{\text{peak area of A in standard solution}}{\text{peak area of IS in standard solution}} \times \text{wt of sample}}$$

where IS is the internal standard. Purity of standard may also be called the assigned assay.

$\prod$ Now calculate the assay of compound *A* from the following information:

Sample solution

Weight of sample	0.1023 g
Peak area of component A	35 312
Peak area of internal standard	33 935

Standard solution

Weight of standard	0.1011 g
Purity of standard	98.9% wt/wt
Peak area of component A	36 004
Peak area of internal standard	34 041

$$\text{Assay of A} = \frac{(35\,312/33\,935) \times 0.1011 \times 98.9}{(36\,004/34\,041) \times 0.1023}$$

$$= \frac{1.0406 \times 0.1011 \times 98.9}{1.0577 \times 0.1023}$$

$$= \quad 96.2\% \text{ wt/wt}$$

Now check the result to make sure it makes sense.

Peak areas for the internal standard in the sample and the standard chromatogram are very similar, so that part is correct.

The response for the compound peak in the sample chromatogram is less than that for the compound peak in the standard chromatogram yet the mass of sample used in the preparation of the sample solution was slightly larger than that used in the standard solution preparation. Thus the assay of the sample must be less than the assigned assay, or purity, of the standard and that is how it has worked out, i.e.

$$96.2\% < 98.9\%$$

Internal standard assays may also be carried out on minor components in a mixture but it is wise to ensure that the magnitudes of the standard and internal standard peaks more accurately reflect the magnitude of the peaks to be determined.

On repetitive injections of the same solution in HRGC it is possible to obtain % relative standard deviations on the assay values of 0.5% or less. It is of more value, however, to carry out the analysis on duplicate or triplicate sample solution preparations since this provides a better assignment of the method precision and the sample homogeneity. These criteria would be established in method validation.

SAQ 7.5b | Why is the internal standard assay method more useful in quantitative analysis?

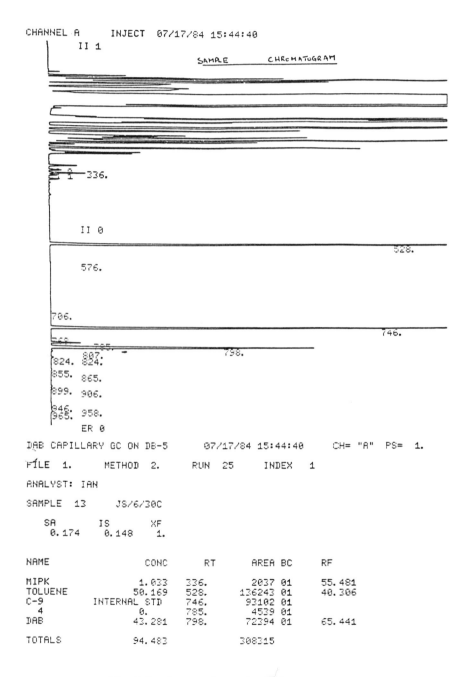

Fig. 7.5a. *Internal standard chromatogram*

Figure 7.5a shows the print-out of the internal standard analysis of a compound called DAB which was prepared as an approximately 50% solution in toluene.

Analysis was carried out on a non-polar capillary using on-column injection. Diethyl ether was used as solvent for this analysis and was obviously not of very good quality.

Toluene eluted at 528 s, the internal calibrant, n-nonane, at 746 s and the product at 798 s. The data printout identifies the analyst, the sample number and the weights of sample and standard sample: also shown are the calibration factors determined from the calibration analysis.

The resultant DAB content was 43.3% wt/wt and the toluene content 50.2% wt/wt. Methyl isobutyl ketone was also determined but, as discussed earlier, it should have been determined with respect to a weaker internal calibrant eluting earlier in the chromatogram.

As you will have noticed, this chromatogram was run many years ago. The degree of sophistication which can be obtained from present day PC based data systems makes it possible to handle and present the data in which ever way one might wish, to provide statistical treatment of data and tabulation of results. However, the format shown is adequate and presents the essential information very clearly.

7.6. METHOD VALIDATION

Most companies and laboratories have their own protocols for method validation. Clearly the extent of method validation should reflect the application of the method and the importance of the product and the significance of results in terms of sales, application or on-going processing.

Generally the criteria to be considered will include repeatability of injection, resolution, peak symmetry, linearity and recoveries. The reproducibility of the method, that is the ability to transfer it to a different column of the same type, to a different chromatograph or a different laboratory, will also be evaluated. Don't become dependent on a method which is 'not transferable'.

Method validation is an expensive activity so it should be carried out to a prescribed plan which makes maximum use of laboratory time and the data generated. Do not embark on a programme until you, as a chromatographer, are convinced that it will stand up to evaluation. Develop the method further if you are in any doubt, aiming for reproducibility and robustness.

SAQ 7.6a | List the parameters you think should be evaluated in a method validation exercise.

Learning Objectives of Part 7

After studying the material in Part 7, you should be able to:

● discuss the sources of error in quantitative gas chromatography;

● describe the peak area normalisation method and appreciate its limitations;

● describe the internal standard method;

● outline the basis of method validation and system suitability testing.

8. Qualitative Analysis

Many years ago when I worked for a leading scientific instrument manufacturer, I recall a visitor coming to see a demonstration of one of the early gas chromatographs. He watched patiently and silently as the recorder pen traced out the peaks as they eluted from the column.

'There you are,' the demonstrator said at the conclusion of the chromatogram, 'what do you think?'

'Is that all?' responded the visitor. The sales manager was taken by surprise. 'What were you expecting to see?' he asked politely. 'Well, all the chromatographs I have seen in the journals and the books have the names of the compound printed alongside the peaks; your instrument does not appear to be able to do that,' came back the reply.

Thirty years on and our visitor still would not be satisfied, or would he?

Gas chromatography began very much as a qualitative technique with most emphasis being placed on separating and possibly identifying components in a mixture rather than determining how much of each was present.

Chromatographers and organic chemists are still primarily interested in *what* is present in the mixture although if you can also tell *how much*, then so much the better.

8.1. RETENTION TIMES

Initially, identification was based upon retention time of the peak of interest compared with the retention time of an authentic compound. This was fairly simple and provided analysis conditions were kept constant, the conclusions were fairly sound. The method was, however, dependent on having samples of the potential compounds of interest. If you synthesised a novel compound then you tended to be 'up the creek' with nothing to compare it with.

SAQ 8.1a What precautions must you take if you wish to use retention times as a means of peak characterisation?

As time went by, chromatographers looked for greater precision in their retention investigations. As a means of circumventing much of the interest variability in retention times, Martin and James adopted the use of relative retention measurement. Relative retention measurement is based on comparing the corrected retention time of the peak of interest with the corrected retention time of an added standard. The use of corrected retention time recognises the observation we made earlier that all components spend the same time in the gas phase and that differences in retention between compounds are due to their relative affinities for the stationary phase.

The corrected or adjusted retention time of a peak is the retention time for the peak, less the retention time of an unretained peak, e.g., an air peak or a butane peak. Butane has little retention on GC columns and has the advantage that it is detectable by FID. This is illustrated in Fig. 8.1a.

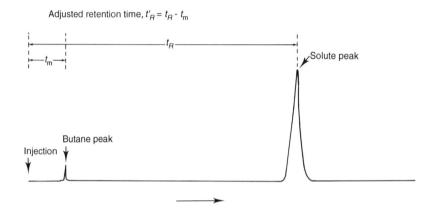

Fig. 8.1a. *Chromatogram showing butane peak*

∏   A major peak on a chromatogram has a retention time of 5.50 mins. Butane elutes in 1 min 45 s. What is the corrected retention time of the peak?

I make it 3 min 45 s. Do you agree? Be certain you know how parts of a minute are presented by your particular data system. Some use seconds, others divide the minute into 100 parts. You can always keep it simple and only use seconds, as the numbers never become unmanageable.

Kovats developed a retention index system based on the relative retentions of an extensive range of organic compounds to other key compounds which he assigned as markers.

Comprehensive tables based on different columns and temperatures were developed, but in my experience their use was never very widely accepted. Perhaps they were really only an interim solution to the problem of identification which was really solved with the introduction of combined GC–MS.

It would be quite wrong to say that retention times are no longer used; perhaps the truth is that we use them more than we realise and almost daily draw conclusions as to the possible identity of unknown components based on their relative retention with respect to compounds we do know.

SAQ 8.1b	What is the advantage of using the relative retention time method?

Much of analytical chemistry is detective work and we must learn to use all the clues which come our way; retention time is one such clue.

SAQ 8.1c	You think you have identified a peak on your chromatogram so you add some of that compound to your mixture; this is called spiking. What would you expect to see on your next chromatogram?

SAQ 8.1c

8.2. SAMPLE ISOLATION FROM GAS CHROMATOGRAPHY

Sample isolation from gas chromatography is really most appropriate to packed column systems, where sufficient material is eluted from the column to make it worthwhile.

Preparative gas chromatographs have appeared on the market from time to time with larger capacity and unfortunately less efficient columns. These systems were not cheap.

∏ Do you think it is essential to use preparative columns for sample isolation?

The answer is no. Packed columns of normal dimensions can be used quite effectively for solute trapping. In many respects they are preferable since their efficiency is generally markedly superior to that of preparative scale columns. The secret is to load the column with increasing quantities of the mixture to be analysed until you reach the stage where resolution of the component you wish to trap is just about to be lost.

It is possible to make a simple system for trapping GC fractions from bits and pieces usually found in laboratory drawers. The essential part is a T-piece with two lengths of stainless steel capillary to which the end of the column can be connected, Fig. 8.2a. With carrier gas flowing through the column, the capillaries are crimped using pliers to allow approximately 10% of the flow to go to the detector connection, with the remainder to go out through the other capillary, through a heated zone in the oven wall, or through a spare injector heater. The outlet capillary should be connected to a short length of tube into

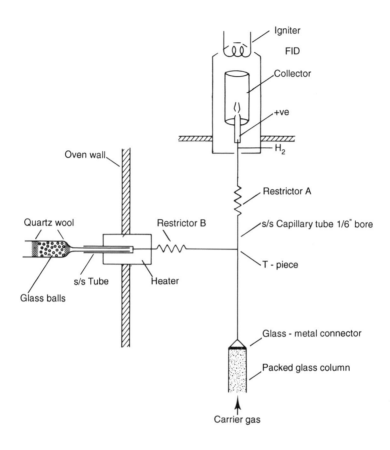

Fig 8.2a. *Schematic representation of a simple laboratory set-up for isolating sample fractions by GC*

which a glass pasteur pipette will neatly fit. This wider tube should also be in the heated zone. The pasteur pipette should be packed with either glass wool or very small glass balls (ballotini) to provide a large surface area to promote condensation but creating very little back pressure or gas hold-up.

It is good practice to leave a pasteur in place all the time and only to replace it with the one in which you want to trap material when the peak is detected on the FID trace.

Additional cooling can be used but, from experience, it does not increase the trapping efficiency significantly. If cooling is too extreme many solutes seem to form 'smokes' which pass straight through the trap.

Usually a number of runs will have to be made before you will visually detect sample condensate in the tube, but even at that level there will probably be sufficient material for high resolution nuclear magnetic resonance or mass spectrometer analysis.

This technique is not very efficient but in my experience it can be a source of small quantities of high-purity material.

SAQ 8.2a   How else may the vaporised solute band exiting the chromatograph be trapped?

8.3. HRGC–MS

The direct combination of high resolution gas chromatography with mass spectrometry has almost become routine, with many inexpensive systems now available, usually PC driven with both the chromatograph and the mass spectrometer controlled from the keyboard.

These systems are based on the simple quadrupole or ion trap analysers and are fitted with modern turbo pumps or diffusion pumps which can easily handle the flows from conventional capillary columns (0.1–0.3 mm internal diameter) without the need for any interface device. The end of the capillary column normally passes right through into the ion source of the spectrometer. Figure 6.5a shows the construction of a combined HRGC–MS system.

To all intents and purposes the mass spectrometer is a GC detector and as such was discussed in some detail in Section 6.5.

∏   Why do you think quadrupole mass analysers are used on bench-top instruments rather than magnetic sector analysers?

If you answered 'cost' you would be in part correct. Quadrupoles are cheaper to make but they are also physically much smaller and place less demands on the pumping system. Although the mass range may be more limited than that of a magnetic analyser, up to 600 amu is quite sufficient for HRGC–MS.

Setting up and running a bench-top HRGC–MS system is well within the capabilities of the average chromatographer with an interest in extending his/her expertise.

It is possibly best to set up the instrument with a non-polar or slightly polar capillary column as a general workhorse. You must be able to handle as much of your work-load as possible without the need to spend time changing columns. You should allow about half a day to change columns although you can reduce this time by cooling the mass spectrometer down overnight.

SAQ 8.3a What is the limiting factor in deciding the diameter of the capillary column which may be used in the mass spectrometer?

A DB-1 or DB-5 column will be suitable for well in excess of 90% of required analyses. There are low bleed columns specifically prepared for GC–MS work and if you can afford them they are worth the extra cost. These columns show significantly less baseline drift during extended temperature programmed runs, reducing the need for baseline subtraction when generating spectra.

Another useful device is one which maintains constant flow through the column during the temperature programme, usually referred to as a constant pressure–constant flow or electronic flow control unit. These devices can reduce the overall analysis time by up to 20%, frequently without loss of resolution, while at the same time significantly reducing baseline drift during the temperature programme. For HRGC–MS I would certainly recommend the flow programming option.

The power in the combined HRGC–MS system lies in the ability to generate mass spectra of each of the components in a complex mixture as they elute from the column. These spectra can either be interpreted from first principles or the system library database can be searched and spectral matches listed. Figure 8.3a shows the library match of the spectrum of the 5.47 min peak on Fig. 6.5b, providing positive identification of the compound.

Fig. 8.3a. (opposite) *Use of a system library database, showing the match of the peak at 5.47 min in Fig. 6.5b*

Library Searched : wiley.l
Quality   : 94
ID   : Benzene, (1-methylpropyl)-

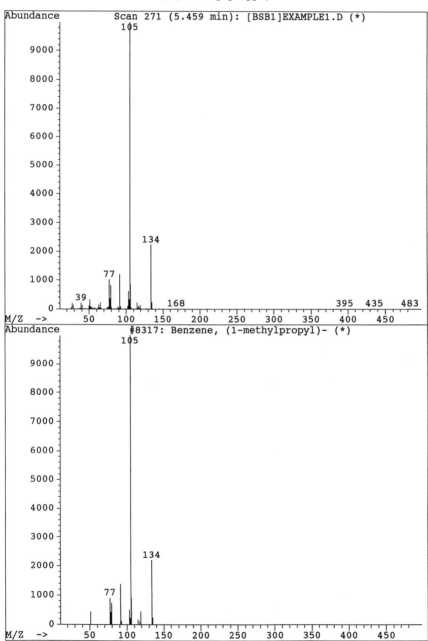

Even if this does not provide positive identification it can often provide clues as to the possible identity or compound type of the eluted peak.

The total ion current (TIC) trace, i.e. the chromatogram obtained from the mass spectrometer, can be processed in the same way as any other detector trace to give quantitative data.

8.4. HRGC–IR

Infrared spectrophotometry as a detection method for HRGC has been discussed in Part 6.6. Again this is a spectroscopic method which will provide component identification or at least structural information about the component eluted from the chromatograph.

Most of the comments in the previous section are again relevant and need not be repeated here.

8.5. HRGC–IR–MS

Here we are getting into the realms of combined techniques.

You may ask why it should be necessary to add IR into a MS system or vice versa.

Ideally, the direct combination of GC with MS or GC with IR spectrophotometry would provide us with positive identification of the components of interest. Unfortunately, we do not live in an ideal world and therefore our systems frequently fail to provide unequivocal answers. This being the case, we are forced to take what useful information we may glean from a number of different techniques and use the information like pieces of a jigsaw to provide a solution.

The most significant advantages of combined systems is that you obtain both spectra simultaneously, thus eliminating the doubt as to whether the mass spectrum and the infrared spectrum actually correspond.

SAQ 8.5a What is the advantage of a combined HRGC–IR–MS system?

SAQ 8.5b Why HRGC–IR–MS and not HRGC–MS–IR?

8.6. HPLC–HRGC–MS

Combining HPLC with HRGC is a relatively recent development in chromatography.

HPLC and HRGC are quite different in terms of what they can achieve. We say that HPLC is a more selective form of chromatography since different chemical types or classes of compounds can be separated one from another by choosing mobile phase compositions and conditions which differentiate between the differing chemical characteristics. Although column efficiency is important we cannot normally run columns longer than 25 cm and therefore the total number of theoretical plates available for the separation may be only a few thousand.

HRGC, on the other hand, can provide us with many tens of thousands or even hundreds of thousands of theoretical plates providing excellent resolution, based on efficiency, but less based on selectivity. Since the carrier gas imparts no influence on the separation, only the stationary phase has any influence on the selectivity of the system.

By combining the two techniques, it is possible to achieve both selectivity and resolution.

The fundamental problem with combining HPLC with HRGC is in dealing with the HPLC mobile phase. Aqueous based mobile phases used in reverse phase HPLC are generally not compatible with capillary columns, particularly if they contain salts and ion pairing reagents. Normal phase HPLC is therefore normally combined with HRGC since the mobile phase is most likely to be a mixture of volatile organic solvents.

Selected parts of the HPLC chromatogram are transferred to a long retention gap connected to the analytical column through either a series of switching valves, or through an automatic cold on-column injector operating in large-volume injection mode. The use of retention gaps is described in Part 5.

It is important that the retention gap is sufficiently long to retain the whole of the transferred fraction. If liquid sample is allowed to reach the analytical column then solute focusing will be degraded or even lost altogether. This limits the volume of mobile phase which can be transferred at any one time. As the mobile phase evaporates, it does

Fig. 8.6a. (opposite) *Analysis of a dichloromethane extract of the aqueous waste liquor stream from the plant scale nitrosation of phenol, using the HPLC–HRGC–MS technique*

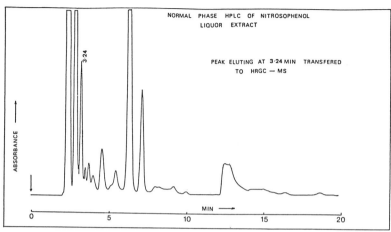

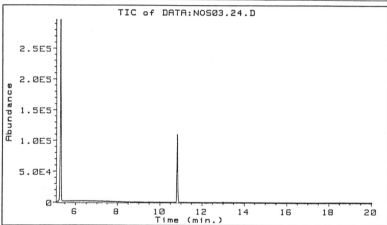

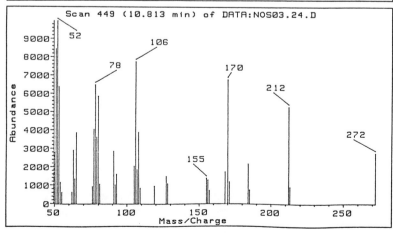

so from the injection end carrying with it and concentrating the solutes as it does so. The solutes move forward until they encounter the stationary phase of the analytical column, Fig. 5.4a, where they focus as a sharp band since the column temperature is only sufficient to volatilise and elute the solvents. The focused components are then separated by temperature programmed elution with identification of the eluted components by mass spectometry.

Figure 8.6a shows the normal phase HPLC chromatogram obtained from a dichloromethane extract of the aqueous waste liquor stream from the plant scale nitrosation of phenol. The total ion current (TIC) trace corresponds to the HRGC–MS analysis of the transferred fraction containing the peak eluting at 3.24 min on the liquid chromatogram and the mass spectrum of the resolved 3.24 min peak eluting at 10.8 min on the gas chromatogram. The early peak on the TIC trace corresponds to the preceding peak on the HPLC chromatogram, which is very large by comparison with the 3.24 min peak.

Clearly, the potential of the procedure is the clean-up efficiency of the LC step, the sensitivity of HRGC–MS and the elimination of material handling.

SAQ 8.6a

> List the problems to be overcome in combining HPLC with HRGC.

SAQ 8.6a

Learning Objectives of Part 8

After studying the material in Part 8, you should be able to:

- discuss identification by retention data and recognise its limitations;

- describe the direct combination of HRGC with mass spectrometry and infrared spectrophotometry;

- discuss the advantages of these combined techniques;

- describe techniques for preparative gas chromatography.

9. Analysis of Less-volatile Samples

9.1. HIGH-TEMPERATURE STATIONARY PHASES

The introduction of bonded phase capillary columns considerably extended the boundaries of gas chromatography, particularly in terms of expanding the range of volatility of samples which might be analysed by the technique.

It has frequently been said in the past that the use of gas chromatography was limited by the volatility of the solute. This was, of course, not strictly accurate, since the limiting factor was more often the volatility of the stationary phase. Bonded non-polar phase capillary columns with upper temperature limits of 325 °C and polar phase columns capable of working up to 230 °C firmly established HRGC and were a very significant advance on packed column systems.

A number of manufacturers now offer variants of the non-polar or slightly polar capillary columns which are rated for use up to 480 °C. More polar phase columns remain limited to about 250 °C.

Π Assuming the problems associated with high-temperature columns can be overcome, what other problems are likely to become more severe as we move into higher-temperature analysis?

Care would need to be taken to avoid sample degradation, particularly at injection. If you need 450 °C to elute your solutes and you are using a vaporising injector, then the solute is going to have to survive a massive thermal shock if you operate the injector at temperatures around or in excess of 400 °C. The situation may be better if you use on-column injection provided you are injecting into the column oven and not into an inter-oven zone.

The temperature-limiting factor with capillary columns is now due more to the instability of the polyimide coating than to the stationary phase. The high-temperature columns are now aluminium clad and, provided rapid temperature cycles which cause the aluminium to become brittle and fracture are avoided, these columns will provide excellent results with remarkably low bleed rates.

With careful use you should be able to operate your capillary columns at temperatures in excess of the manufacturer's suggested limit. This will only be true if you have run your column in over a period, gradually increasing the upper operating temperature, and you have taken the precautions discussed earlier in this book.

Whether a sample may be analysed by HRGC or not always proves a challenge. You will find that many compounds which 'could not possibly be analysed by GC' can be analysed very successfully with very little difficulty.

9.2. DERIVATISATION

Chromatographers fall into two categories when it comes to derivatisation. There are those who adopt it with enthusiasm and those who consider it as a last-resort measure.

I must be included in the latter group. If derivatisation appears as the only hope for gas chromatographic analysis, then I look for alternative chromatographic or spectroscopic methods.

Derivatisation may be an excellent way of making a compound sufficiently volatile for GC analysis, but in the case of precise quantitative analysis it introduces additional levels of method performance which must be validated. For a one-off sample it is appropriate, but for repetitive on-going analysis then it is probably prudent to investigate alternatives.

SAQ 9.2a

> Would you expect the following analyses to require special treatment because of the low volatility of the sample? Circle the correct answer.
>
> (i)   The fatty acids in a sample of soap.   Y/N
>
> (ii) A sample of diesel oil.   Y/N
>
> (iii) The phenols used as raw materials for
>   preparing phenolic resins.   Y/N
>
> (iv) A phenolic resin.   Y/N
>
> (v)   A light machine oil.   Y/N

The objective of derivatisation is to increase the volatility of the analyte. Compounds which exhibit intermolecular hydrogen bonding, e.g. alcohols, phenols, carboxylic acids and amines, usually have higher boiling points than might be anticipated on the basis of their molecular weights.

The potential for hydrogen bonding may also influence the manner in which the analyte interacts with the chromatographic column. Hydrogen bonding may occur between the analyte and polar groups on the column and this will give rise to peak tailing.

Clearly if the hydrogen bonding could be eliminated then the compound would become more volatile, making it more suitable for GC analysis, and peak tailing would be reduced in the chromatographic analysis.

∏ Can intermolecular hydrogen bonding be reduced simply by vaporising the compound?

The answer is yes. Benzoic acid exhibits very strong intermolecular hydrogen bonding which is only slightly dissociated in a non-polar solvent. However, at 150 °C in the vapour state, hydrogen bonding totally disappears.

There are a number of comprehensive texts on derivatisation which provides guidelines for reagent selection and preparative methods. These are listed in the Bibliography.

The most common derivatisation methods involve acetylation of alcohols, phenols and amines using acetic anhydride in the presence of an acid catalyst.

$$ArOH + (CH_3CO)_2O \longrightarrow CH_3CO_2Ar + CH_3COOH$$

Triflouroacetyl derivatives are also useful for low-level analysis by electron capture detection. Trifluoroacetic anhydride of triflouroacetylimidazole are used as derivatising reagents.

Carboxylic acids are usually converted to their methyl esters either by using diazomethane, boron trifluoride/methanol complex or methanol and a catalytic amount of acid, as preferred.

High-percentage conversions are possible, but for quantitative analysis this is one parameter which must be under control.

Trimethylsilyl derivatives are popular and are relatively easy to prepare, i.e. by simply reacting the analyte with one of the proprietary reagents in a screw-capped vial at room temperature or about 60 °C.

Three common reagents are listed below:

TMS
Trimethylchlorosilane (Me_3SiCl)

HMDS
Hexamethyldisilazane ($Me_3Si-N=N-SiMe_3$)

BSA
N,O-bistrimethylsilylacetamide ($CH_3-C(OSiMe_3)=N-SiMe_3$)

A typical reaction is as follows:

$$2\,R\!-\!OH \ + \ \underset{\displaystyle CH_3\!-\!\overset{\textstyle O-SiMe_3}{\overset{|}{C}}\!=\!N\!-\!SiMe_3}{} \ \longrightarrow \ 2\,R\!-\!OSiMe_3 \ + \ CH_3CONH_2$$

Silylating reagents are rarely pure and frequently generate volatile reaction by-products which produce interfering peaks on the sample chromatogram. It is therefore prudent to run a blank containing all the materials apart from the analyte to ensure separation and positive identification of the analyte peak in the derivatised sample solution.

SAQ 9.2b	Complete the following equation: How would you carry out this reaction?

9.3. PYROLYSIS GAS CHROMATOGRAPHY

Pyrolysis gas chromatography is a further technique which might be used to provide gas chromatographic data from involatile materials. The compound is decomposed by heat to give volatile components which are then separated chromatographically to give a fingerprint chromatogram. If mass spectrometric detection is used, then identification of the decomposition products may provide important structural information about the original material.

Pyrolysis chromatography has been used in the analysis of polymers and plastics.

Three criteria must be satisfied for pyrolysis chromatography data to be meaningful:

(1) The decomposition products should be characteristic of the product.

(2) The reaction and the conditions under which the decomposition takes place must be reproducible.

(3) The process must be quick and easy to perform.

Satisfying (2) and (3) comes down to equipment design. For example, if the temperature changes then the nature and the profile of the decomposition produced may also change. If the decomposition process is too slow then, unless measures are taken to focus the volatile components at the front of the analytical column, a broad unrepresentative initial sample band may result which will degrade the potential resolution.

Different types of pyrolysis attachments have been introduced over the years, e.g. furnaces, flash heaters and Curie-point pyrolysers, Fig. 9.3a.

The Curie-point pyrolyser is probably the most widely accepted device used in pyrolysis gas chromatography at present.

(a)

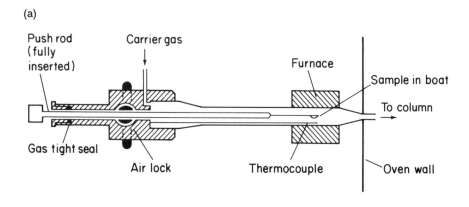

(b)

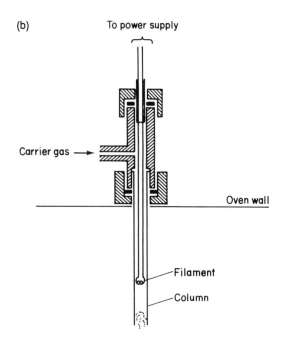

Fig. 9.3a. *Pyrolysis devices: (a) furnace pyrolyser; (b) filament pyrolyser; (c) Curie-point pyrolyser*

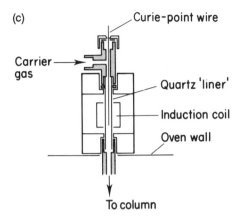

Fig. 9.3a. (*continued*)

Learning Objectives of Part 9

After studying the material in Part 9 you should be able to:

● discuss the range of high-temperature columns;

● discuss the reasons for derivatisation and the appropriateness of the alternative methods;

● describe pyrolysis gas chromatography.

10. Environmental Analysis Systems

10.1. INTRODUCTION

Environmental analysis is a large and expanding area of analytical chemistry at the present time and it is true to say that gas chromatography plays a very significant role in this activity.

Whether monitoring the immediate environment of the workplace, factory emissions or waste streams, etc., or the wider environment, many of the pollutants which concern us are organic compounds which may be readily analysed by gas chromatography with flame ionisation detection. Indeed, gas chromatography coupled to mass spectrometry is an ideal combination to aid not only the detection but also the identification of pollutants.

The most common problem associated with environmental analysis is detectability. The human nose is very sensitive and although we may be able to detect an odour it may be very much more difficult to detect and identify the components which make up that odour, analytically using gas chromatography.

The first step in an analysis scheme is therefore usually some means of increasing the concentration of the volatile pollutant(s) present in the headspace of the sample or in the atmosphere being examined. This may be accomplished by pumping the atmosphere through a solid phase adsorbant, a suitable filter medium or simply by condensation in a low-temperature trap. Analysis is then carried out on the concentrated sample.

Not all environmental pollutants are particularly volatile at normal temperatures and pressures and in many cases they may be present in complex matrices. Here some form of solvent extraction may be necessary to liberate the analyte for analysis.

SAQ 10.1a | List the problems likely to be encountered by the analytical chemist carrying out environmental analysis.

10.2. SOLVENT EXTRACTION

In the case of aqueous solutions or solid samples it may be necessary to extract the pollutants with an organic solvent to provide the necessary concentration enhancement. Recent research has shown that superficial fluids may also be used as extractants.

Supercritical carbon dioxide has been used to extract compounds from difficult samples and of course there is an additional advantage that the carbon dioxide can readily be removed by evaporation to leave the concentrated analyte. Further, by introducing a modifier into the CO_2 and by controlling the density of the supercritical fluid it may be possible to be selective in the extraction and pull out only the compounds of interest. A lot of development work is going on into this process at present. In addition, the use of CO_2 eliminates the costs of organic solvent use and disposal.

The major problem associated with organic solvent extraction of low-level analytes is the level of solvent impurities and other matrix associated compounds which may co-extract and interfere with the analysis. If solvent evaporation is an additional necessary step in the process then this frequently aggravates the problem. Large-volume injection into capillary columns with solute focusing, as described in Part 5, can effectively increase the sensitivity of the method and ease the final stage of the analysis.

Construction of calibration graphs is an essential part of low-level component analysis but it is probably less time consuming to establish the analysis conditions using a solution containing the compounds of interest at about ten times the anticipated sample level. In that way you can build up your expertise without any doubts about what you are seeing on the chromatograms. When you are well practised and have confidence in the method and the chromatography, then proceed with the calibration analysis.

Successful analysis of low-level organic compounds requires a great deal of attention to detail and rigorous cleaning of equipment used in the procedures. Some companies have 'clean' laboratories specifically set up for this type of analysis.

SAQ 10.2a | Suggest alternative approaches for the analysis of low-level volatile organic compounds in water.

SAQ 10.2a

SAQ 10.2b | Suggest alternative approaches for the analysis of low-level organic compounds contained in a soil sample.

SAQ 10.2b

10.3.   ON-LINE FOCUSING

For HPLC analysis, aqueous pollutants may be directly focused on a suitable pre-column or adsorbant filter simply by passing the water sample through the column or the filter. Partition will favour the bonded phase and the concentrated organics can then be eluted into an analytical column with the mobile phase containing a significant proportion of organic modifier. This also applies for SFC, the only qualification being that the bulk of the aqueous phase needs to be removed from the pre-column before switching in the supercritical mobile phase. Passing aqueous solutions through a GC column is not really a practical proposition. However, focusing of volatile organic compounds in both packed and capillary columns is quite feasible.

For example, an aqueous phase may be scavenged by a gas flow and the volatiles reconcentrated on a suitable adsorbant. The trap is then heated to desorb the focused solutes onto the analytical column. This is the basis of purge and trap methods. Alternatively, the volatiles may be focused on-column in a capillary column as described in the next section.

10.4.   HEADSPACE ANALYSIS IN HRGC

Headspace methods can be categorised into static and dynamic methods. In the static case, a fixed volume of the headspace atmosphere is taken, usually by gas syringe, and injected into the chromatographic system. Provided the injection volume is not too large and that there is a degree of solute focusing on the column, then reasonable results may be obtained. This technique is most useful for intermediate volatility solutes, which may be forced into the headspace by incubating the sample solution in a closed vial at say 40–60 °C prior to removal of the sample. In the case of very volatile solutes, the components may not be significantly retarded by the stationary phase and broad initial bands will result, leading to poor chromatography.

Dynamic systems, in which the headspace is sampled over a period of time by passing a scavenging flow over or through an odorous sample or simply by pumping the gaseous atmosphere into a trap, also depends on partition or adsorption of the solutes on the trapping column in the first instance. Such a system is used on personal air samplers carried by those working in potentially hazardous working environments. The sampling traps may contain Tenax, charcoal or other suitable adsorbent. The trapped volatiles may be desorbed using a small quantity of solvent or by thermal desorption onto the analytical column using a minimum volume of gas.

It is possible to focus volatile components very efficiently if cryogenic cooling is used at the beginning of the column. Grob established a rule that if there was a temperature difference of 80 °C between the compound boiling point and the elution temperature then the solute will be sharply focused. The volatile components may be transferred into the cooled column through an on-column injector or through a split/splitless injector operating in splitless mode.

Figure 10.4a shows a very simple arrangement which may be used for qualitative analysis of volatiles in solid or liquid samples.

Figure 10.4b shows the headspace chromatogram resulting from the dynamic transfer and focusing of the headspace of an aqueous solution of cinnamon oil using this set-up with a splitless injector. It is interesting to note the range of volatiles which may be detected from an ambient temperature headspace and the fact that minor components in the sample appear enhanced in the headspace environment.

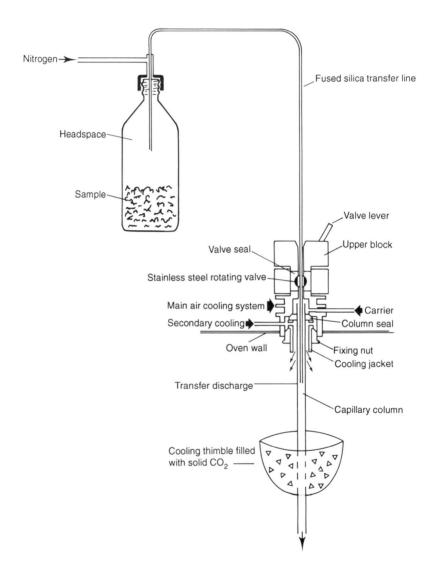

Fig. 10.4a. *A simple headspace sampling system*

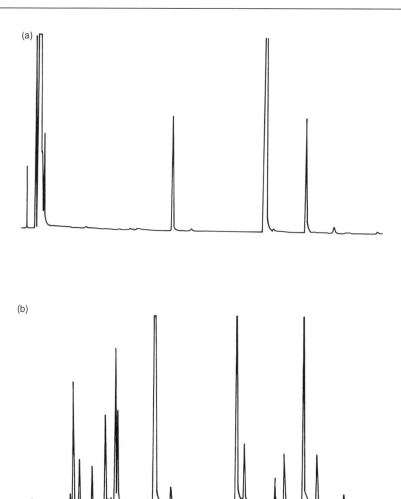

Fig. 10.4b. *Analysis of cinnamon oil: (a) normal injection through the split/splitless injector (ethyl acetate solution); (b) headspace analysis through the split/splitless injector (dilute aqueous solution)*

10.5. PURGE AND TRAP SYSTEMS

Purge and trap methods are becoming quite common in environmental analysis and instrument companies are now offering instruments which will sample a stream or waterway, purge out the volatiles, concentrate them in a trap and then thermally desorb the components into the chromatographic column for automatic analysis. Indeed the complete system may be automated for 24 hour–7 day operation.

In order to ensure that the concentrated solute band is refocused at the beginning of the analytical column some systems include cryogenic cooling of the initial section of the column during the desorption phase.

Figure 10.5a shows the flow diagram of a commercial system for

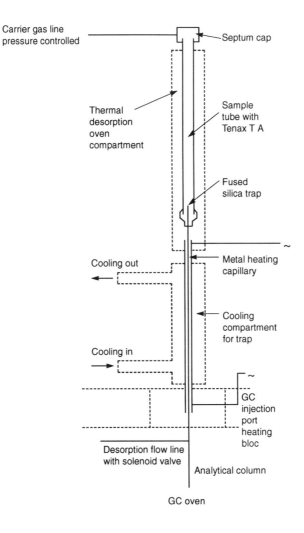

Fig. 10.5a. *Flow diagram of a commercially available purge and trap system*

analysis by thermal desorption of a sample adsorbed on a Tenax trap or by desorption of a sample adsorbed on a suitable support contained in the glass liner. The volatiles desorbed from the trap are refocused on a length of cooled capillary column which is then heated rapidly to transfer the solutes as a sharp band into the analytical column oven. Liquid nitrogen is used to cool the capillary trap.

Equipment is now available to which a number of purge tubes can be fitted and the system programmed to process them automatically.

It seems likely that environmental analysis will be the major area of development in the application of gas chromatography in the coming years.

Learning Objectives of Part 10

After studying the material in Part 10 you should be able to:

- describe the basis of specialised equipment for headspace analysis and thermal desorption analysis which may be used in environmental analysis.

Self-assessment Questions and Responses

SAQ 1a

> List the differences between packed and capillary columns.

Response

Packed Columns	Capillary Columns
Glass, 0.5–3 m	Fused silica, 10–60 m
Typically 4 mm i.d.	Typically 0.2–0.3 mm i.d.
Contain solid support	No solid support
Liquid phase on solid support	Liquid phase on column wall
Flow rate 20–50 ml/min	Flow rate 1–2 ml/min
Simple sample injection	Specialist injectors required
~5 K Theoretical plates	50–100 K Theoretical plates
Broad peaks	Narrow peaks
Lower trace sensitivity	Higher trace sensitivity
High sample loading	Low sample loading
Good for simple mixtures	Ideal for complex mixtures
Less skill required	More skill required

SAQ 2.1a | Which gases are normally used for gas chromatography?

Response

Nitrogen and helium. In the UK, helium is expensive compared to nitrogen and hence less widely used. For GC–MS, helium must be used.

SAQ 2.1b | Could the following gases be used for gas chromatography and if so do they have any particular limitations or advantages; hydrogen, argon and carbon dioxide?

Response

All three can be used as carrier gases. Hydrogen, due to its smallest molecular mass is the most suitable gas for capillary column gas chromatography. However, the risk of explosion puts most workers off using it.

Argon was widely used in the past and was the essential carrier gas in the days of the argon ionisation detector. It was still used extensively until fairly recently. Being of larger molecular weight it would not really be suitable for HRGC.

Carbon dioxide could be used as a carrier in gaseous form but it is now extensively used as a mobile phase in supercritical fluid chromatography (SFC). In SFC the operating conditions are above the critical temperature and pressure and the resulting mobile phase has solubility characteristics more akin to a liquid than a gas. SFC can be considered as dense gas chromatography. The advantage of SFC is the ability to analyse more polar materials than would normally be possible by GC.

SAQ 2.3a

> Why must different injection systems be used in capillary column gas chromatography, i.e. HRGC?

Response

Capillary column gas chromatography works on a very much smaller scale than packed GC. To ensure that a representative sample of the solutes to be analysed is placed on the capillary column as a narrow band, a number of injection devices have been developed. In split/splitless injection the sample is vaporised and a portion passed into the capillary column where it may be focused as a narrow band prior to elution and separation. With on-column injection a suitably designed syringe places the sample injection right into the column. This requires a precision-engineered injection port to provide alignment of the column and the syringe needle.

SAQ 2.8a

> List six items which should be checked when starting up a gas chromatograph.

Response

(1) That the carrier gas has been turned on and is flowing through the column.

(2) That the electrical power to all the units has been switched on.

(3) That the temperature setting for the column is appropriate to the phase and the application.

(4) That injection and detector heaters have been switched on and set correctly and that they are warming up.

(5) That the detector flame is alight, if it is an FID.

(6) That the system responds to an injection of solvent.

SAQ 3.2a Glass and stainless steel tubing may be used in the manufacture of packed columns for gas chromatography. List two advantages and one disadvantage for each.

Response

The following are lists of the 'pros' and 'cons'. For packed column work, glass columns are hard to beat and if handled with some care breakage is rare. Occasionally, if the column dimensions are not correct then the column will be strained when mounted in the oven and this may lead to failure.

Glass
Advantages: Chemically inert
Thermal stability
Very low adsorptive capacity
Transparency
Disadvantage: Fragility

Stainless Steel
Advantages: Chemically inert
Thermal stability
Mechanical strength
Disadvantages: Opaque
Occasional catalytic decomposition

SAQ 3.4a List three respects in which diatomaceous earth falls short of the ideal as a support for gas chromatography.

Response

(1) It retains some adsorptive capacity.

(2) It is physically fragile and so breaks down into smaller particles resulting in a wider particle size range.

(3) The particles are irregular in shape.

(4) It has deep pores which fill with liquid stationary phase. This can result in band broadening as the solute molecules may take longer to diffuse into the stationary phase and out again to the liquid phase surface. (Remember — 'no long tunnels'.)

SAQ 3.7a.
> By applying the principle of *like dissolves like*, for each of the following pairs of compounds, suggest which one will be most soluble in the stated solvent.
> (a) naphthalene and 1-naphthol in benzene
> (b) benzophenone and diphenylmethane in propanone
> (c) methanol and butan-1-ol in water.

Response

(a) The correct answer is naphthalene, which would be more soluble in benzene since it, too, is an aromatic hydrocarbon while 1-naphthol is a phenol.

(b) The correct answer is benzophenone, which, like propanone, is a ketone. Diphenylmethane is a hydrocarbon.

(c) The correct answer is methanol. Both methanol and butan-1-ol are alcohols and share the OH functional group with water. However, they differ in the length of the hydrocarbon chain, the longer chain of butanol making it more hydrophobic than methanol.

SAQ 3.7b | By applying the principle of *like dissolves like* for each of the following pairs of compounds, suggest which one will elute first from a column of the stated stationary phase.

(a) methylbenzene (bp = 100 °C) and ethyl 2-methylpropanoate (bp = 110 °C) on squalane (a saturated hydrocarbon)

(b) butan-1-ol (bp = 116 °C) and 4-methylpentan-2-one (bp = 117 °C) on glycerol

(c) hexane (bp = 68 °C) and 1-methylethyl methanoate (bp = 68 °C) on PEG-S (polyethylene glycol succinate).

Response

You have probably noticed that each pair of compounds has essentially the same boiling point so there will be no significant differences between their volatilities. Thus separation will again have to be based upon solubility differences.

(a) Ethyl 2-methylpropanoate. The stationary phase is an alkane, so it will be a better solvent for the hydrocarbon, methylbenzene, than for the ester, ethyl 2-methylpropanoate. This will cause the methylbenzene to be retained more and so ethyl 2-methyl-propanoate will be eluted first.

(b) 4-Methylpentan-2-one. The stationary phase is a polyhydric alcohol which will dissolve the alcohol, butan-1-ol, more than the ketone. The ketone therefore elutes more rapidly.

(c) Hexane. The stationary phase is a polyester and as such will dissolve the ester, 1-methylethyl methanoate, more than the hydrocarbon, hexane. Hexane will therefore elute first.

SAQ 3.7c

Place the following stationary phases in order of increasing polarity:

polyethylene glycol succinate (PEG-S),

polyethylene glycol — relative molecular mass 400 (PEG 400),

polyethylene glycol — relative molecular mass 20000 (PEG 20M),

hexadecane,

tritolyl phosphate,

polypropylene glycol adipate (PPGA).

Response

Hexadecane < tritolyl phosphate < PEG 20M < PEG 400 < PPGA < PEG-S.

PEG-S is the most polar, consisting mainly of ester carbonyl groups; PPGA is less polar, the carbonyl groups being 'diluted' by the extra methylene groups; PEG 400, the polyether, is the next most polar, being more polar than PEG 20M because it has a higher proportion of terminal hydroxyl groups because of its lower relative molecular mass. Tritolyl phosphate is less polar again since the three aryl groups 'dilute' the polar phosphate group and the alkane. Hexadecane is least polar.

SAQ 3.7d

Benzene, cyclohexane and ethanol are to be separated by GLC. Given that they all boil between 70 °C and 80 °C, indicate by circling either T for true or F for false whether you agree with either of the following statements.

(1) On a squalane stationary phase at 70 °C the order of elution would be:

ethanol, followed by benzene, followed by cyclohexane.

T / F

(2) On a polyethylene glycol succinate (PEG-S) stationary phase at 70 °C, ethanol would elute after both benzene and cyclohexane.

T / F

You might like to consider what would happen if PEG 400 was used as a stationary phase instead.

Response

(1) If your response was T, well done. You obviously realised that with such similar boiling points, the separation will be solely due to differences in solubility.
If your answer was F, sorry. We have something to sort out here. With such similar boiling points, the separation will be solely due to differences in solubility. Squalane will not form hydrogen bonds or dipole attractions. The only forces left to encourage solubility are dispersion forces which are strongest between similar molecules, i.e. the two alkanes, squalane and cyclohexane, and weakest between dissimilar molecules, i.e. squalane and ethanol. The order of elution in the statement is therefore correct.

(2) If your answer was T, we are going to have to work on it, because it was wrong. This is because benzene and ethanol elute together,

after cyclohexane. The weak hydrogen bonding between ethanol and the polyester PEG-S is swamped by the big dipole–dipole attractions between the carbonyl of the polyester and the hydroxyl of the alcohol. This is matched by the large dipole-induced dipole attraction between the carbonyl of the polyester and the dipole it induces in benzene by polarising its π electrons. Hence their similar solubility.

If your answer was F, well done.

You might like to consider what would happen if PEG 400 were used as stationary phase instead.

Answer to supplementary question:

If you suggested that ethanol would elute *after* benzene, you were right. You realised that ethanol would now form much stronger hydrogen bonds and that benzene would form much weaker dipole-induced dipole attractions as a result of the smaller dipole and greater hydrogen bonding ability of PEG-400. Ethanol would now be more soluble than benzene and so elute after it.

SAQ 3.7e

> For the three mixtures below, select in each case the answer (a), (b) or (c) which you think is the correct order in which the named components of the mixture will elute on the given stationary phase.
>
> (1) Cyclohexane (bp = 81 °C) and cyclohexene (bp = 83 °C) on dinonyl phthalate (DNP).
>
>   (a) More or less together.
>   (b) Cyclohexane then cyclohexene.
>   (c) Cyclohexene then cyclohexane.
>
> (2) Methoxybenzene (anisole, bp = 154 °C) and 1-methylethylbenzene (cumene, bp = 152 °C) on polyethylene glycol (PEG 400).
>
>   (a) More or less together.
>   (b) Anisole then cumene.
>   (c) Cumene then anisole.
>
> (3) Hexane (bp = 68 °C), 1-methyl-1-(1-methyl-ethoxy) ethane (diisopropyl ether) (bp = 68 °C) and propan-2-ol (bp = 83 °C), on polyethylene glycol succinate (PEG-S).
>
>   (a) Hexane, diisopropyl ether, then propan-2-ol.
>   (b) Diisopropyl ether, hexane, then propan-2-ol.
>   (c) Hexane, propan-2-ol, then diisopropyl ether.

Response

(1) Correct answer = (b). DNP is slightly polar, in spite of its very high proportion of alkyl and aryl groups. The carbonyl groups can induce a small dipole in cyclohexene by polarising the π-electrons. This will increase cyclohexene's solubility in the stationary phase, which, together with its slightly lower volatility, will lead to it eluting after the completely unpolarisable cyclohexane.

If you answered (a) I assume that you missed the polarisability of the

π-electrons, but if you answered (c) I am at a loss to know what you were thinking. If you have read the discussion above and cannot see where you went wrong, then you probably need to discuss this with your tutor.

(2) Correct answer = (c). PEG 400 has a high proportion of hydroxyl groups which can act as proton donors in hydrogen bonding with the oxygen atom of the ether. PEG 400 does not have a strong enough dipole to induce a dipole in the aryl rings of either anisole or cumene by polarising the π-electrons. Consequently, the ether is more soluble than the hydrocarbon. Combined with its slightly higher boiling point this means that the ether migrates more slowly. Incidentally, had we chosen PEG-S as the stationary phase, because it is a much poorer proton donor in hydrogen bonding and also has a much greater dipole, we would not have separated the mixture nearly as well.

If you answered (a) you probably missed the difference in hydrogen bonding ability. Again, there is no obvious reason for answering (b), so if you did this you ought to find someone to discuss it with, unless the above comments have made you see where you were going wrong.

(3) Correct answer = (a). The propan-2-ol is less volatile and has a greater solubility in the stationary phase due to its dipole and hydrogen bonding ability (as a proton donor). The ether has a lesser dipole and will not hydrogen bond with PEG-S as neither of them has a hydrogen atom. Hexane has no dipole and no hydrogen bonding ability.

SAQ 3.8a	Explain, in the space below, why gases are more likely to be analysed by GSC than GLC.

Response

Your explanation should have included mention of the fact that gases are not sufficiently soluble in most liquids at 50 °C and above for them to be retained significantly by GLC so that very long columns and sub-ambient temperatures would have to be used, whereas gases are strongly adsorbed on many solids. Consequently retention would be reasonable on short columns above room temperature. Separation will be possible. You should also have mentioned the greater selectivity of adsorption due to the rigid geometry of adsorbents as well as molecular sieving, both of which will be necessary to separate gases which may have very little difference in polarity.

Your answer may have been less detailed, but if you did not get substantially the above answer, perhaps you should re-read 'Solid Stationary Phases' in Section 3.8 of this Unit.

SAQ 3.8b	If you were asked to determine the concentration of carbon monoxide and carbon dioxide in a boiler flue, which of the following stationary phases would you use? (1) To determine carbon monoxide. (2) To determine carbon dioxide. Stationary phases: Alumina, silica gel, graphitised carbon black, Linde molecular sieve, porous polystyrene (Porapak).

Response

You may have noticed that all the stationary phases suggested to you are adsorbents. Liquid stationary phases are generally of little use for separating non-hydrocarbon gases unless sub-ambient temperatures and very long columns are used.

(1) A Linde molecular sieve is probably the best phase for separating

carbon monoxide from the other gases in air (see Section 3.3 of this Unit). Porous polymers are not generally very good for separating permanent gases, being, like graphitised carbon black, better for organic liquids. Alumina and silica gel do not separate carbon monoxide from oxygen or nitrogen.

(2) Silica gel would be needed to separate carbon dioxide. You could not use a molecular sieve because carbon dioxide is irreversibly adsorbed, which makes the analysis slightly tedious — you have to inject onto two separate columns.

SAQ 4.1a | Golay's expectations were not realised in his initial experiments in capillary chromatography. What do you think the inadequacies of the injection systems and detectors were that are referred to in the text?

Response

At the time of Golay's initial experiments, commercial packed column gas chromatographs were in their infancy. Microlitre syringes were not available and home-made glass capillaries were the normal method of injection. Indeed it was difficult to get small enough volume injections for packed column systems, thus injection into capillaries were orders of magnitude more difficult. In my first month in 1960 working as a technician with gas chromatography, I do not think I managed to get a peak on scale, and I was almost ready to look for a new job.

Only katharometers and gas density balances were available at that time and the internal volume of these detectors was of similar order to that of the column, so it was quite possible to lose any separation obtained in the column within the voids of the detector. In addition, these detectors were very much less sensitive than argon ionisation detectors and flame ionisation detectors which were introduced in the early 1960s.

SAQ 4.2a | Describe, with the aid of a labelled diagram, the construction of a capillary column which you might buy today from a laboratory supplier. How does it differ from a column manufactured in the 1970s and what are the advantages of the modern product?

Response

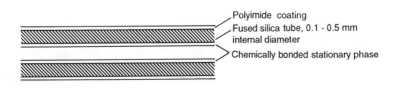

Polyimide coating
Fused silica tube, 0.1 - 0.5 mm internal diameter
Chemically bonded stationary phase

Original capillary columns were based on glass, steel or plastic tubing coated with stationary phase by deposition on the internal surface. Glass columns were fragile and required skilled handling, steel columns suffered from adsorption and plastic columns could not stand temperature. Early columns were restricted to generally non-polar phases as the polar phases tended not to wet the inner surface adequately to provide an even film.

Modern columns are based on very high purity fused silica, coated with polyimide on the outside to protect them from atmospheric oxidation and with the stationary phase chemically bonded to the internal column surface.

The method of construction, the high purity of the materials used and the chemically bonded phase all serve to extend the operating temperature range of both apolar and polar phase columns.

Columns are robust, flexible, and with practice easy to handle and are capable of providing several years of service if properly used and maintained.

SAQ 4.3a	Arrange the following bonded phase columns in order of increasing polarity: DB-17, DB-5, Ultra-2, DB-1701, HP-225, CP-Sil 5, SBP-1, HP-2, DB-WAX.

Response

CP-Sil 5, SPB-1	100% methyl
DB-5, Ultra-2, HP-2	5% phenyl/95% methyl
DB-1701	14% cyanopropylphenyl/86% methyl
HP-225	50% cyanopropylphenyl/50% methyl
DB-WAX, HP-20M	polyethylene glycol

This was not a trick question, as several of the phases were equivalent. One of the great difficulties for new chromatographers is the wide assortment of designations for essentially the same column type. There are at least 15 equivalents of DB-1. It is a pity the manufacturers cannot get their act together!

There is no need to remember them all, but do use the catalogues as an information source.

You will probably find that columns differ more in physical characteristics than in chromatographic performance.

Note that as you might expect, the operating temperature range of the column decreases with increasing polarity.

SAQ 4.4a

> List four basic precautions to take when installing a capillary column in a gas chromatograph.

Response

A number of possibilities that you may have chosen are listed, they are all relevant to successful installation.

- Ferrules and fittings are placed onto the column before cutting the ends.

- Column ends are cleanly cut and free from fragments or shattering and are clean.

- Connections are not overtightened.

- The column is hanging freely and not stressed at couplings or round supports. I know you can tie fused silica columns in knots but do not do it!

- The column is protected from radiant heat in the oven.

- Carrier gas is flowing at the appropriate rate.

- Check for leaks but NOT with soap solution; use ethyl acetate.

- Allow carrier gas to flow through the column for 45 min before switching on the heaters.

- Ensure that the column oven and injector and detector heater set temperatures are compatible with the column. It would not be the first column to be over-cooked before use if you get it wrong!

SAQ 4.6a

> Which of the following gases would be suitable as carrier gases in HRGC? Circle the appropriate response: yes (y), no (n).
>
> | | |
> |---|---|
> | Argon | y/n |
> | Hydrogen | y/n |
> | Carbon dioxide | y/n |
> | Air | y/n |
> | Nitrogen | y/n |
> | Helium | y/n |
> | Xenon | y/n |
> | Oxygen | y/n |

Response

Obvious choices would be hydrogen, helium and nitrogen although you may have reservations about hydrogen from a safety point of view. Air and oxygen would be bad news for your column. Argon and xenon are inert gases and could certainly be used but would be more appropriate to packed column gas chromatography. Argon was used in the days of the argon ionisation detector and remained in use for a long time later when people switched to flame ionisation detection.

Carbon dioxide is used as a carrier gas in supercritical fluid chromatography (SFC) in both packed and capillary columns. CO_2

will act as a carrier in the form of a liquid, a gas or a supercritical fluid and has rather better solubility characteristics than the traditional carrier gases. SFC may be thought of as dense gas chromatography.

SAQ 4.7a

> The actual time spent by all solute molecules in the gas phase as they pass through the column is the same. True or False?

Response

If you said false you have still not got to grips with your understanding of the chromatographic process be it packed column, capillary or even liquid chromatography.

The differences in retention time in an isothermal chromatogram are due to differences in the distribution constants K_d.

The actual time spent by all the molecules in the gas phase *is* the same but the time spent in the liquid (or stationary) phase increases with K_d.

SAQ 4.7b

> With reference to the relationship below, predict how the retention time will change on changing the parameters.
>
> $$t'_R = \frac{C_S}{C_M} \times \frac{2d_f}{r} \times \frac{L}{\bar{u}}$$
>
> (1) Decreasing the film thickness, d_f.
> (2) Increasing the internal diameter.
> (3) Reducing the linear gas velocity, $\bar{u}$.
> (4) Increasing the column length, L.
> (5) Increasing the temperature.

Response

This was fairly straightforward I trust.

(1) Decreasing the film thickness reduces the retention time. Remember that reducing the film thickness will also reduce the

sample load handling capacity of the column and you may need to reduce the injection volume or the sample solution concentration to avoid peak overloading.

(2) Increasing the internal diameter, say to 0.5 mm, i.e. 'Megabore', will significantly reduce the retention time since it will be necessary for the solute molecule to spend more time in the gas phase between contacts with the liquid film.

(3) Reducing the linear gas velocity will increase the retention times.

(4) Increasing the column length will obviously increase the retention time if we do not do anything else. Remember that increasing the column length by a factor of two doubles the number of theoretical plates available but resolution is only increased by factor of $\sqrt{n}$.

(5) On the face of it, temperature does not appear to feature in the equation, yet there are only two ways to change the relationship C_S/C_M and that is by either changing the character of the stationary phase or by changing the temperature such that the value of the distribution constant changes. Thus, increasing the temperature will reduce the concentration in the stationary phase, increasing the concentration in the mobile phase and thereby reducing the retention time.

SAQ 4.7c	Suggest a column and operating conditions which will provide you with a good chance of successfully analysing more than 70% of your samples at the first attempt.

Response

From long experience I have found that the following choice of column and conditions satisfies a very large proportion of my separation needs.

- Column: J & W DB-5, or equivalent, 30 m $\times$ 0.25 mm i.d. with a 0.25 μm film thickness.

- Temperature: 70 °C isothermal for 2 min, programme at 16 °C/min to 300 °C, then 10 minute final isothermal period.

- Sample solvent: AnalaR ethyl acetate.

Try it — you can improve on it later, but if someone needs an analysis quickly you will be able to provide it and become very popular with your boss!

SAQ 4.7d

> The phantom chromatographer has struck during the night and stolen your favourite DB-1 column. This was a 30 m × 0.25 mm i.d. column with an efficiency at optimum gas velocity of 112 000 theoretical plates. Urgent analysis is required and the only DB-1 column you have available is 15 m × 0.32 mm i.d. Calculate by how much your resolution is likely to be reduced by using this column. Suggest an alternative column for your analysis.

Response

Your calculation should have progressed as follows:

We shall assume optimum gas velocities throughout.

> 30 m × 0.25 mm gave 112 000 plates, therefore:
> 15 m × 0.25 mm should yield 56 000 plates.

Since efficiency is inversely proportional to the internal diameter and your 15 m column is 0.32 mm i.d., the efficiency will be reduced to:

$$56\ 000 \times 0.25/0.32 = 37\ 750 \text{ plates}$$

Resolution changes with the square root of efficiency, thus the factor for the reduction in resolution will be:

$$\sqrt{43\,750}\,/\,\sqrt{112\,000}\,=\,0.62$$

This is a significant loss in resolution.

Rather than accept this loss of resolution, I would use an equivalent dimension DB-5 column, 30 m × 0.25 mm i.d. There may be slight changes in relative retention times and analysis may take a little longer, but I would expect the resulting chromatogram to be recognisable when compared with that obtained from DB-1.

SAQ 5.1a | Which of the following statements regarding injection into capillary columns are correct? Circle YES or NO.

(a) Normal GC syringes cannot be used against the high pressures used in HRGC.   YES/NO

(b) HRGC columns have limited capacity requiring lower sample loading.   YES/NO

(c) Conventional syringes will not fit into a capillary column.   YES/NO

(d) Samples must be completely vaporised before they enter the capillary column.   YES/NO

(e) The two most common injection techniques in HRGC are split/splitless vaporisation and cold on-column injection.   YES/NO

(f) All injectors are non-discriminating and give good quantitative results in HRGC.   YES/NO

Response

(a) NO. Inlet pressures in HRGC are very modest, typically 14 psi (100 kPa). Syringes can operate against very high pressures as the cross-sectional area of the plunger is so small.

(b) YES. The limited sampling handling capacity was the driving force behind the development of the vaporising injectors.

(c) YES and NO. Metal needle syringes will not fit into 0.2–0.3 mm i.d. columns but fused silica needles will, hence on-column injection.

(d) NO. Samples may be in the liquid or the gaseous state on the column. In on-column they will certainly be liquids and even in vaporising injection the solvent and the solutes may recondense on the column wall when they enter the column oven.

(e) YES.

(f) Unfortunately NO. As will be discussed in Section 6.2 sample discrimination remains a potential problem in split/splitless injection.

SAQ 5.2a | There are three ways in which gas may leave a vaporising injector; what are they and what is their function?

Response

(i) As carrier gas onto the analytical column carrying a small portion of the vaporised sample.

(ii) As split flow removing the bulk of the vaporised sample. The split flow determines the time the sample vapour cloud remains in the chamber and in certain circumstances this may influence the initial band width on the analytical column.

(iii) As the septum purge flow. This flow is generally quite small and ensures that no septum bleed components or sample residues deposited on the septum can reach the analytical column.

SAQ 5.2b | List three ways in which the use of a split/splitless injector may result in the resulting chromatogram not being representative of the injected sample.

Response

(i) If the sample is thermally labile and likely to degrade then response is likely to be reduced and decomposition products, if they are volatile, may appear on the resulting chromatogram, further confusing the analysis.

(ii) As the sample vaporises in the injector, not all the components in the mixture vaporise at the same rate. This may result in the sample entering the column not being truly representative of the whole, i.e. sample discrimination may take place. In addition, if the column is not fitted correctly or the syringe needle is too long, then droplets of liquid sample may enter the column as an unrepresentative sample, possibly overloading the column.

(iii) Sample residues may build up on the glass linear in the chamber. These residues may act as adsorptive sites or promote sample decomposition.

SAQ 5.3a | List three advantages of cold on-column injection.

Response

(i)   Direct injection into the cool column. No thermal shock to the solutes.

(ii) The total sample is injected onto the column. Provided the injection part of the column or the retention gap is free from residues there is little possibility of solute discrimination.

(iii) On-column injection ensures that the repeatability of the chromatography and the data should match the precision of syringe injection.

SAQ 5.3b | May both fused silica and metal needle syringes be used in on-column injection?

Response

Yes. Generally fused silica needles must be used for direct injection into 0.2–0.3 mm i.d. columns. Where a 0.5 mm i.d. retention gap or a 0.5 mm i.d. 'Megabore' column is being used then a metal needle syringe may be used.

SAQ 5.3c | Why is it important to depress the syringe plunger smartly at the moment of injection?

Response

It is important to ensure that the column of liquid detaches from the needle tip and is not allowed to form a liquid deposit between the wall of the column and the needle.

SAQ 5.4a | What is the function of a retention gap?

Response

The function of a retention gap was initially to provide a means of on-column injection into capillary columns using automatic injectors fitted with metal syringe needles. Subsequently, it was realised that retention gaps provided a means of focusing solutes from large-volume injections at the front of the analytical column and this led on to their use in combining HPLC with HRGC.

SAQ 5.6a | What is the advantage of combining HPLC with HRGC?

Response

The chromatographic processes in HPLC and HRGC are to some extent quite different. HPLC separates on the basis of selectivity while HRGC separates on the basis of efficiency. Combining the two techniques brings together the best of both worlds and certainly where the combination also includes mass spectrometric detection then the ability to transfer an HPLC fraction for further high resolution separation and MS detection and identification is a powerful combination.

SAQ 6.1a

In Section 6.1 we listed the desirable characteristics of a detector as follows:

High sensitivity
Universal response
Wide linear dynamic range
Good stability
Low background noise
Reliability
Ease of operation.

Which of these characteristics do you consider to be most important in:

(a) quantitative analysis;
(b) environmental analysis;
(c) in a laboratory with a very heavy workload?

Response

In (a) good stability and a wide linear dynamic range would be most important. Good stability can ease the calibration workload and of course a wide linear range will improve quantitative accuracy.

Environmental analysis, (b), frequently requires high sensitivity for trace component analysis. Stability is probably more important than linearity since trace analysis is often carried out using spiked additions over very limited ranges. Low background noise will help to lower levels of detection.

In the busy laboratory, (c), reliability and ease of operation may be most important in ensuring sample throughput.

SAQ 6.1b

Which GC detector do you consider to be in most common use today?

Response

The flame ionisation detector is certainly the most commonly used detector in gas chromatography at present. In the future, with the reducing cost of mass spectrometer hardware and software, the mass spectrometer may rival the FID to a greater extent than at present.

SAQ 6.2a   List three attributes of the flame ionisation detector
and one limitation.

Response

Attributes (1) Almost universal response to organic compounds;
only halogenated compounds give anomalous
responses.

(2) Wide linear dynamic range makes the FID an
excellent detector for accurate quantitative analysis.

(3) High sensitivity, ideal for the detection of trace
impurities in complex mixtures.

Limitation   The flame ionisation detector does not respond to
water, the permanent gases, carbon monoxide,
carbon dioxide, formaldehyde and formic acid.

SAQ 6.4a   Which carrier gas would you choose for each of the
following analyses, assuming that you only have a
gas chromatograph fitted with a hot wire detector?

(a) The determination of traces of hydrogen in air.

(b) The separation of propanone and propan-2-ol.

Response

(a) Nitrogen or argon would be appropriate since these would
enhance the thermal conductivity difference. Helium would not
be satisfactory because it has a similar thermal conductivity to
hydrogen and therefore sensitivity would be poor.

(b) Nitrogen would probably be satisfactory but if high sensitivity
were required for trace analysis, then hydrogen or helium would
be more suitable.

SAQ 6.5a | What are the most significant advantages of the mass spectrometer over other detectors?

Response

The mass spectrometer can produce mass spectra of each component eluting from the column in addition to quantitative data. This can provide the chemist with rapid identification of components in mixtures. The mass spectrometer is both universal in response and selective upon demand.

SAQ 6.6a | Classify the detectors we have considered in terms of universal response, specificity, sensitivity and ease of use, allowing 1 for excellent, 2 for good and 3 for poor.

Response

Detector	Universal	Specific	Sensitive	Ease of use
FID	1	3	1	1
ECD	3	1	1	3
HWD	1	3	2	1
MS	1	1	1	1
IR	2	1	2	2

SAQ 7.1a

> What do you consider to be the most important characteristics of the chromatographic system if it is to produce accurate quantitative analysis?

Response

Linearity of response over a wide dynamic range for a large range of organic compounds, and baseline separation of the components to be quantified are prerequisites for accurate quantitative analysis. Universal response would be ideal but it is not essential since analytes can be determined by comparison with known amounts of known-purity samples of the same compound.

SAQ 7.2a

> List four possible methods of quantifying a chromatogram in the order you think they might have been introduced in GLC.

Response

(1) Peak height measurement.

(2) Peak area measurement by cutting and weighing.

(3) Peak area measurement by triangulation.

(4) Peak area measurement by digital integration.

SAQ 7.2b

> Comment on the suitability of the methods you listed in SAQ 7.2a.

Response

Peak height measurement is quite good if we are comparing the same component peak on successive chromatograms. It is of little value in assessing the relative composition of a mixture of compounds, particularly in the case of an isothermal chromatogram where peak width increases with time and therefore the peak height decreases.

Peak area measurement by cutting and weighing can be quite accurate provided you take certain precautions, i.e. run the chart at a high speed to give good sized peaks, avoid handling the 'peaks', cut out the peaks accurately and dry them in a desiccator prior to weighing.

Triangulation is quick and simple. If a fast chart speed is used then measurement of peak width will be more accurate. However, you will always have to approximate and omit part of the front and rear of your peak in the construction, and this will give rise to errors.

Digital integration is usually superior to other methods provided the integration parameters are set correctly.

SAQ 7.2c	Under what circumstances will integration always involve a degree of approximation?

Response

Integration of unresolved peaks will always involve a degree of approximation depending upon how the method handles the lack of resolution.

SAQ 7.2d	What is the solution to the problem in SAQ 7.2c?

Response

Improving the chromatographic separation to give baseline resolution of the peaks will improve the accuracy of the integration process. This may involve a change in the analysis conditions or even the use of a different stationary phase.

SAQ 7.3a   Why is variability of injection volume more of a problem in GC than in HPLC?

Response

Most injections into gas chromatographs are made by syringe, whereas in HPLC loop injection valves are normally used. The HPLC injection method simply involves filling the sample loop with solution and turning the valve handle to incorporate the sample loop into the analytical flow line. Manual injection into gas chromatographs using a microlitre syringe requires a much higher degree of manipulative skill to achieve acceptable repeatability of very much smaller sample volumes, i.e. 0.1–1.0 μl, compared with 10–20 μl. Automatic injection usually provides better repeatability than manual injection, which is not surprising, since the system will carry out precisely the same operation time after time.

SAQ 7.3b   Which analysis and data handling method overcomes the problem of injection variability in gas chromatography?

Response

The internal standard method.

SAQ 7.5a   What are the limitations of the peak area normalised method?

Response

Retained components or peaks which are not integrated are not taken into account in the calculation of the composition of the mixture or the purity of the major component. This can give rise to incorrect results. Remember the method only area-normalises the peaks which the integration system detects.

SAQ 7.5b | Why is the internal standard assay method more useful in quantitative analysis?

Response

(i) Samples are analysed with respect to a standard sample of known purity.

(ii) Samples for analysis are accurately weighed, (or the volume accurately measured) and the results calculated on a wt/wt (or vol/vol) basis.

(iii) The use of an internal standard overcomes the problem of variability of injection volume.

SAQ 7.6a | List the parameters you think should be evaluated in a method validation exercise.

Response

Resolution

Efficiency

Peak symmetry

Repeatability of injection

Robustness

Recovery of spiked additions

Reproducibility of the method on analysis of a number of preparations of analysis solutions of the same sample.

There may be others you wish to include.

| SAQ 8.1a | What precautions must you take if you wish to use retention times as a means of peak characterisation? |

Response

When using retention times it is important to ensure that conditions remain constant and that the injection and run-start procedure is always carried out in precisely the same manner.

Column flow-rate, temperature, temperature programme conditions, injection and data system start routine must not be changed between injections of samples and standards. It is also important to adjust retention times such that they can be accurately measured. If retention times are reasonable and you wish to make measurements on the chromatogram, then run at a faster than normal chart speed.

| SAQ 8.1b | What is the advantage of using the relative retention time method? |

Response

Since the relative retention time is based on the ratio of the corrected retention times of the component under investigation and that of a chosen standard, the value will be less influenced by small changes in the analysis conditions. It is important, however, that the retention times should be similar in magnitude.

| SAQ 8.1c | You think you have identified a peak on your chromatogram so you add some of that compound to your mixture; this is called spiking. What would you expect to see on your next chromatogram? |

Response

We would expect to see a peak at the same retention time as before; the peak should show no sign of distortion which might indicate that the added component was different, and of course the peak should be enlarged in proportion to the amount of 'spike' we added.

If we were to run the mixture on a different stationary phase and still see only a single peak for the spiked sample, then that would be additional useful confirmation.

SAQ 8.2a | How else may the vaporised solute band exiting the chromatograph be trapped?

Response

A short packed column containing a suitable adsorbent may be fitted to the outlet and the solute band subsequently removed with solvent or by heating. I have, on occasion, packed a small quantity of potassium bromide (KBr) into the narrow section of the pasteur pipette and once the component has been trapped, removed the KBr and pressed it into a micro-disc for infrared analysis.

Another possibility is to pass the column eluent through a small volume of solvent, say chloroform or carbon tetrachloride for infrared solution analysis or, alternatively, deuterochloroform might be used and the resulting solution examined by NMR spectroscopy. The one problem with this approach is the loss of solvent to the gas stream.

SAQ 8.3a | What is the limiting factor in deciding the diameter of the capillary column which may be used in the mass spectrometer?

Response

The ability of the mass spectrometer's pumping system to remove the carrier gas and retain a vacuum in the order of 10^{-6} torr determines the maximum diameter of the column.

Turbo pumps and the diffusion pumps built into bench-top instruments can handle up to approximately 2 ml/min. This flow-rate range is appropriate for 0.1–0.3 mm i.d. columns but not for 0.5 mm 'Megabore' columns.

SAQ 8.5a | What is the advantage of a combined HRGC–IR–MS system?

Response

The ability to generate both infrared and mass spectra, simultaneously, of each component as it elutes from the column.

SAQ 8.5b | Why HRGC–IR–MS and not HRGC–MS–IR?

Response

The mass spectrometer must come last in the sequence since it is a destructive technique and requires high vacuum.

Splitting the outlet from the HRGC equipment and sending half to the mass spectrometer and the other half to the IR spectrophotometer would be less efficient as this would half the potential sensitivity of both detection systems.

SAQ 8.6a | List the problems to be overcome in combining HPLC with HRGC.

Response

(1) The nature of the HPLC mobile phase.

(2) Establishing a suitable normal phase HPLC method.

(3) The transfer system.

(4) Establishing the practical transfer volume.

(5) Removing the mobile phase from the HRGC while focusing the solutes.

SAQ 9.2a

> Would you expect the following analyses to require special treatment because of the low volatility of the sample? Circle the correct answer.
>
> (i) The fatty acids in a sample of soap.   Y/N
>
> (ii) A sample of diesel oil.   Y/N
>
> (iii) The phenols used as raw materials for preparing phenolic resins.   Y/N
>
> (iv) A phenolic resin.   Y/N
>
> (v) A light machine oil.   Y/N

Response

(i) Correct answer, YES. Soap is a mixture of the sodium salts of stearic acid ($C_{17}H_{35}CO_2H$) and palmitic acid ($C_{15}H_{31}CO_2H$) plus other additives to make them, and us, feel good and smell good. The acid salts must be converted to their esters to make them sufficiently volatile for gas chromatographic analysis.

(ii) Correct answer, NO. Diesel oil is relatively volatile and can be analysed on non-polar columns.

(iii) Correct answer, NO. Phenolic compounds used as raw materials in the manufacture of resins can be readily analysed by gas chromatography.

(iv) Correct answer, YES. Phenolic resins are solids and as such would not be volatile. The best chance of obtaining a chromatogram representative of the resin would be by pyrolysis gas chromatography.

(v) Correct answer, probably NO, although you may use a short thin film column to obtain a reasonable chromatogram, since light machine oil is rather involatile. Supercritical fluid chromatography (SFC) might well be a better option.

SAQ 9.2b. | Complete the following equation:

$$\text{(C}_6\text{H}_5)\text{OH} + \text{CH}_3-\underset{\underset{\text{O}-\text{SiMe}_3}{|}}{\text{C}}=\text{N}-\text{SiMe}_3 \longrightarrow \ ?$$

How would you carry out this reaction?

Response

$$2\ \text{(C}_6\text{H}_5)\text{OH} + \text{CH}_3-\underset{\underset{\text{O}-\text{SiMe}_3}{|}}{\text{C}}=\text{N}-\text{SiMe}_3 \longrightarrow 2\ \text{(C}_6\text{H}_5)\text{O}-\text{SiMe}_3 + \text{CH}_3-\text{CONH}_2$$

The silylating reagent should be added in excess to the sample in a screw-capped vial. A small amount of pyridine may be added as a catalyst. This may be heated to 60 °C for a few minutes to complete the reaction.

SAQ 10.1a | List the problems likely to be encountered by the analytical chemist carrying out environmental analysis.

Response

The following are some of the problems to be overcome in environmental analysis:

(i) Low-level solutes requiring concentration enhancement prior to analysis.

(ii) Solutes dispersed in matrices not amenable to gas chromatography, e.g. water or soil samples, etc.

(iii) Extraction solvents which contain interfering impurities.

(iv) Need for on-column focusing of large-volume injections of volatile solutes to maintain chromatographic performance.

(v) Large numbers of samples for analysis.

SAQ 10.2a | Suggest alternative approaches for the analysis of low-level volatile organic compounds in water.

Response

(i) Extract the solutes with a non-interfering organic solvent using a solvent/water ratio which will give concentration enhancement. Evaporate the solvent, provided that the solutes are not lost in the evaporation process, and inject a large volume of the concentrated solution into an HRGC system equipped for large-volume injection and fitted with either an FID or a mass spectrometer.

(ii) Place the sample in a suitable purge and trap system and transfer the volatiles to a cryogenically cooled capillary trap for focusing and subsequent transfer to the analytical column.

SAQ 10.2b | Suggest alternative approaches for the analysis of low-level organic compounds contained in a soil sample.

Response

(i) The soil sample may be shaken with a suitable extraction solvent, the solution filtered to remove particulates, concentrated if possible to enhance the concentration of the solute in the solution and analysed by large-volume injection onto a capillary column.

(ii) The soil sample may be placed in a supercritical fluid extraction tube and extracted with either carbon dioxide alone or carbon dioxide modified with methanol. The density of the supercritical fluid should be adjusted to optimise the extraction. The extracted solutes may then be dissolved in a minimum of organic solvent prior to injection into the gas chromatograph. This process would eliminate most of the problems encountered in conventional solvent extraction processes, i.e. solvent impurities, solute loss during evaporation of solvent, etc.

Units of Measurement

For historical reasons a number of different units of measurement have evolved to express quantity of the same thing. In the 1960s, many international scientific bodies recommended the standardisation of names and symbols and the adoption universally of a coherent set of units — the SI units (Système Internationale d'Unités) — based on the definition of five basic units: metre (m); kilogram (kg); second (s); ampere (A); mole (mol); and candela (cd).

The earlier literature references and some of the older text books naturally use the older units. Even now many practising scientists have not adopted the SI unit as their working unit. It is, therefore, necessary to know of the older units and be able to interconvert with SI units.

In this series of texts SI units are used as standard practice. However, in areas of activity where their use has not become general practice, for example biologically based laboratories, earlier defined units are used. This is explained in the study guide to each unit.

Table 1 shows some symbols and abbreviations commonly used in analytical chemistry, while Table 2 shows some of the alternative methods for expressing the values of physical quantities and their relationship to the values in SI units. In addition, Table 3 lists prefixes for SI units and Table 4 shows the recommended values of a selection of physical constants.

More details and definition of other units may be found in D.H. Whiffen, *Manual of Symbols and Terminology for Physicochemical Quantities and Units*, Pergamon Press, 1979.

Table 1   *Symbols and Abbreviations Commonly Used in Analytical*
Chemistry

Å	Angstrom
$A_r(X)$	relative atomic mass of X
A	ampere
E or U	energy
G	Gibbs free energy (function)
H	enthalpy
J	joule
K	kelvin (273.15 + t °C)
K	equilibrium constant (with subscripts p, c, therm, etc.)
K_a, K_b	acid and base ionisation constants
$M_r(X)$	relative molecular mass of X
N	newton (SI unit of force)
P	total pressure
s	standard deviation
T	temperature/K
V	volume
V	volt (J A^{-1} s^{-1})
$a, a(A)$	activity, activity of A
c	concentration/mol dm^{-3}
e	electron
g	gram
i	current
s	second
t	temperature/ °C
bp	boiling point
fp	freezing point
mp	melting point
≈	approximately equal to
<	less than
>	greater than
e, exp(x)	exponential of x
In x	natural logarithm of x; In x = 2.303 log x
log x	common logarithm of x to base 10

Table 2 *Summary of Alternative Methods of Expressing Physical Quantities*

(1) **Mass (SI unit : kg)**

$$g = 10^{-3}\,kg$$
$$mg = 10^{-3}\,g = 10^{-6}\,kg$$
$$\mu g = 10^{-6}\,g = 10^{-9}\,kg$$

(2) **Length (SI unit : m)**

$$cm = 10^{-2}\,m$$
$$\text{Å} = 10^{-10}\,m$$
$$nm = 10^{-9}\,m = 10\,\text{Å}$$
$$pml = 10^{-12}\,m = 10^{-2}\,\text{Å}$$

(3) **Volume (SI unit : m³)**

$$l = dm^3 = 10^{-3}\,m^3$$
$$ml = cm^3 = 10^{-6}\,m^3$$
$$\mu l = 10^{-3}\,cm^3$$

(4) **Concentration (SI unit : mol m⁻³)**

$$M = mol\,l^{-1} = mol\,dm^{-3} = 10^3\,mol\,m^{-3}$$
$$mg\,l^{-1} = \mu g\,cm^{-3} = ppm = 10^{-3}\,g\,dm^{-3}$$
$$\mu g\,g^{-1} = ppm = 10^{-6}\,g\,g^{-1}$$
$$ng\,cm^{-3} = 10^{-6}\,g\,dm^{-3}$$
$$ng\,dm^{-3} = pg\,cm^{-3}$$
$$pg\,g^{-1} = ppb = 10^{-12}\,g\,g^{-1}$$
$$mg\% = 10^{-2}\,g\,dm^{-3}$$
$$\mu g\% = 10^{-5}\,g\,dm^{-3}$$

(5) **Pressure (SI unit : N m⁻² = kg m⁻¹ s⁻²)**

$$Pa = N\,m^{-2}$$
$$atmos = 101\,325\,N\,m^{-2}$$
$$bar = 10^5\,N\,m^{-2}$$
$$torr = mmHg = 133.322\,N\,m^{-2}$$

(6) **Energy (SI unit : J = kg m² s⁻²)**

$$cal = 4.184\,J$$
$$erg = 10^{-7}\,J$$
$$eV = 1.602 \times 10^{-19}\,J$$

Table 3 *Prefixes for SI Units*

Fraction	Prefix	Symbol
10^{-1}	deci	d
10^{-2}	centi	c
10^{-3}	milli	m
10^{-6}	micro	μ
10^{-9}	nano	n
10^{-12}	pico	p
10^{-15}	femto	f
10^{-18}	atto	a

Multiple	Prefix	Symbol
10	deka	da
10^2	hecto	h
10^3	kilo	k
10^6	mega	M
10^9	giga	G
10^{12}	tera	T
10^{15}	peta	P
10^{18}	exa	E

Table 4 *Recommended Values of Physical Constants*

Physical constant	Symbol	Value
acceleration due to gravity	g	9.81 m s^{-2}
Avogadro constant	N_A	$6.022\,05 \times 10^{23} \text{ mol}^{-1}$
Boltzmann constant	k	$1.380\,66 \times 10^{-23} \text{ J K}^{-1}$
charge-to-mass ratio	e/m	$1.758\,796 \times 10^{11} \text{ C kg}^{-1}$
electronic charge	e	$1.602\,19 \times 10^{-19} \text{ C}$
Faraday constant	F	$9.648\,46 \times 10^4 \text{ C mol}^{-1}$
gas constant	R	$8.314 \text{ J K}^{-1} \text{ mol}^{-1}$
'ice-point' temperature	T_{ice}	273.150 K, exactly
molar volume of ideal gas (stp)	V_m	$2.241\,38 \times 10^{-2} \text{ m}^3 \text{ mol}^{-1}$
permittivity of a vacuum	ϵ_o	$8.854\,188 \times 10^{-12} \text{ kg}^{-1}$ $\text{m}^{-3} \text{ s}^4 \text{ A}^2 \text{ (F m}^{-1})$
Planck constant	h	$6.626\,2 \times 10^{-34} \text{ J s}$
standard atmosphere pressure	p	$101\,325 \text{ N m}^{-2}$, exactly
atomic mass unit (amu)	m_u	$1.660\,566 \times 10^{-27} \text{ kg}$
speed of light in a vacuum	c	$2.997\,925 \times 10^8 \text{ m s}^{-1}$

Index